Industrial Catalysis:
Chemistry and Mechanism

Industrial Catalysis:
Chemistry and Mechanism

James D. Burrington

Imperial College Press

Published by

Imperial College Press
57 Shelton Street
Covent Garden
London WC2H 9HE

Distributed by

World Scientific Publishing Co. Pte. Ltd.
5 Toh Tuck Link, Singapore 596224
USA office: 27 Warren Street, Suite 401-402, Hackensack, NJ 07601
UK office: 57 Shelton Street, Covent Garden, London WC2H 9HE

Library of Congress Cataloging-in-Publication Data
Names: Burrington, James D., 1951–
Title: Industrial catalysis : chemistry and mechanism / James D. Burrington
   (The Lubrizol Corporation, USA).
Description: New Jersey : Imperial College Press, 2016.
Identifiers: LCCN 2015043857| ISBN 9781783268979 (hc : alk. paper) |
   ISBN 9781783268986 (sc : alk. paper)
Subjects: LCSH: Catalysis--Textbooks. | Organometallic chemistry--Textbooks. |
   Chemistry, Technical--Textbooks.
Classification: LCC QD505 .B88 2016 | DDC 660/.2995--dc23
LC record available at http://lccn.loc.gov/2015043857

British Library Cataloguing-in-Publication Data
A catalogue record for this book is available from the British Library.

Desk Editors: Kalpana Bharanikumar/Catharina Weijman

Typeset by Stallion Press
Email: enquiries@stallionpress.com

To
Rebecca

Contents

Foreword   XV

1. An Introduction to Industrial Catalysis   **1**

  1.1 Historical Background   1

      1.1.1 Early Catalysis Examples   2

      1.1.2 Highlights of Catalysis Technology   2

  1.2 Catalysis Definitions   6

  1.3 Kinetics and Catalysis   7

  1.4 Thermodynamics   8

      1.4.3 Thermodynamic Limitation   10

  1.5 Key Concepts   11

      1.5.1 Activity: Batch Measures   11

      1.5.2 Activity: Space Velocity   12

      1.5.3 Space Velocity Examples   13

      1.5.4 Activity: Turnover Rate   14

      1.5.5 Turnover Example   15

      1.5.6 Selectivity   17

      1.5.7 Selectivity Example   18

1.5.8 Selectivity Example 2: Formation of Two Products	19
1.5.9 Catalyst Life	21
1.6 Catalytic Mechanism	21
1.6.1 Representing Mechanisms: "Electron Pushing"	22
1.6.2 Representing Mechanisms: "Name Reaction" Examples	23
1.6.3 Elementary Catalytic Reaction Steps	23
1.6.4 Representing Catalytic Mechanisms: The "Tolman Formalism"	25
1.6.5 Catalytic Mechanism Example: Hydroformylation of Olefins	25
1.7 Homogeneous Versus Heterogeneous Catalysis	26
1.7.1 Homogeneous Catalysis Example	28
1.7.2 Heterogeneous Catalysis Example	29
1.7.3 Advantages of Homogeneous Catalysis	30
1.7.4 Advantages of Heterogeneous Catalysis	31
1.8 Major Industrial Catalytic Process Types	32
1.8.1 Industrial Examples	33
1.9 Problems	35
1.10 Answers to Problems	38
2. Acid Catalysis	**43**
2.1 Basic Concepts	44
2.1.1 An Acid Catalyst Periodic Table	44
2.1.2 pKa Acidity Scale	45
2.1.3 Ho Acidity Scale for Superacids	46
2.1.4 Superacid Example: Phosphotungstic Acid	47
2.1.5 Bronsted Versus Lewis Acidity	48
2.1.6 Some Common Solid Acid Catalysts	49
2.1.7 Silica–Alumina SiO_2/Al_2O_3 Solid Acids	50
2.1.8 Zeolite Synthesis	52

2.1.9   Zeolite Acid Sites   52

2.1.10   Bronsted and Lewis Acidity Measurements
         in Solids   53

2.1.11   Mechanism of Acid Catalyzed Aromatic Reactions   55

2.1.12   Thermodynamics   56

2.2 Elements of Acid Catalyst Design   57

2.2.1   Active Site Distribution   57

2.2.2   Zeolites   58

2.2.3   Other Si/Al Frameworks   59

2.2.4   Shape Selectivity   60

2.2.5   Surface Activity Effects   63

2.2.6   Surface Activity: Small Crystallite Example   64

2.2.7   Heat Treatment — Phase Formation   66

2.2.8   Common Catalyst Supports   68

2.2.9   Catalyst Supports: Hydrophobic Versus
         Hydrophillic Surfaces   68

2.3 Major Industrial Processes — Refinery   69

2.3.1   Catalytic Cracking   69

2.3.2   Disproportionation of Aromatics   71

2.3.3   Isobutane/isobutylene Alkylation (Octane)   72

2.3.4   Methyl–tert–butyl ether Synthesis   73

2.3.5   Methanol-to-Gasoline (MTG)   74

2.4 Major Industrial Processes — Chemical   76

2.4.1   Benzene Ethylation   76

2.4.2   Cumene   76

2.4.3   Beckman Isomerization   77

2.4.4   Esterification   79

2.5 Trends in Acid Catalysis   79

2.6 Problems   81

2.7 Answers to Problems   82

3. Oxidation Catalysis   **83**

3.1 Concepts   84

    3.1.1 Oxidation Catalyst Periodic Table   84

    3.1.2 Oxidation Thermodynamics   85

    3.1.3 Recognizing Organic Oxidation States   85

    3.1.4 Oxidation Catalysis Concepts — Homogeneous   90

    3.1.5 Oxidation Catalysis Concepts — Heterogeneous   90

    3.1.6 Elementary Reaction Steps   91

    3.1.7 Homogeneous Oxidation Catalysis   92

    3.1.8 Hetereogeous Oxidation   95

3.2 Elements of Oxidation Catalyst Design   98

    3.2.1 Metals Substitution   98

    3.2.2 Active Site Isolation   99

    3.2.3 Surface Versus Bulk Composition   101

    3.2.4 Promoters   102

    3.2.5 Oxidation Catalyst Synthesis   102

3.3 Major Industrial Processes   104

    3.3.1 Epoxidation — Styrene Monomer Propylene Oxide Process   104

    3.3.2 Acrylonitrile   105

    3.3.3 Vinyl Acetate/Wacker Oxidation   105

    3.3.4 Cyclohexanol/One — Cyclohexane Oxidation   107

    3.3.5 Adipic Acid/Nylon 6,6   108

    3.3.6 Maleic Anhydride   109

    3.3.7 Aromatic Oxidation — Terephthalic Acid   110

    3.3.8 Phthalic Anhydride   111

    3.3.9 Aromatic Oxidation — Cumene to Phenol/Acetone   111

3.4 Problems   113

3.5 Answers to Problems   116

4. Polymerization Catalysis **119**

4.1 Concepts 120

    4.1.1 Polymerization Catalyst Periodic Table 120

    4.1.2 Thermodynamics 122

    4.1.3 Polymer Molecular Weight 122

    4.1.4 Polymerization Reaction Mechanisms 125

    4.1.5 Polymerization Mechanisms: Elementary Steps 125

    4.1.6 Anionic Polymerization 127

    4.1.7 Cationic Polymerization 129

    4.1.8 Metal Oxide/Coordination Catalysis 130

    4.1.9 Coordination Catalysis: Ziegler–Natta 131

    4.1.10 Coordination Catalysis: Metallocene 132

    4.1.11 Free Radical Polymerization 133

    4.1.12 Living Polymerization 135

    4.1.13 Living Polymerization: TEMPO Initiator 136

4.2 Polymerization Catalyst Design 137

    4.2.1 Metals Substitution: PP Isomers 138

    4.2.2 V-Based Polypropylene Catalysts: Syndiotactic 139

    4.2.3 Ti-Based PP Catalysts: Isotactic 140

    4.2.4 Ligand Effect: Constrained Geometry Metallocenes 141

    4.2.5 Isotactic from Cp* Catalyst 142

    4.2.6 Non-Coordinating Anions 143

    4.2.7 Effect of Solvents: Butadiene Example 145

    4.2.8 Ziegler–Natta Catalyst Synthesis:
        Particle Microreactor 146

    4.2.9 Single Versus Multiple Site Polyolefin Catalysts 147

4.3 Major Industrial Processes 149

    4.3.1 Polyethylene and Isotactic Polypropylene:
        $TiCl_4/AlEt_2Cl/MgCl_2$ 149

    4.3.2 Ethylene/Propylene Copolymers (EPC) 150

4.3.3   Higher Olefins: Shell Higher Olefin Process (SHOP)   153

4.3.4   Styrene/Butadiene Emulsion Polymerization (SBR)   154

4.3.5   Polyethoxylates (Anionic Polymerization)   156

4.3.6   Polyisobutylene   156

4.4 Future Catalysts: Late Transition Metals
for Branched Polyethylene   158

4.4.1   Branched Polyethylene   159

4.4.2   Branching Mechanism   160

4.5 Problems   161

4.6 Answers to Problems   166

5. **Reduction/Hydrogenation Catalysis**   **169**

5.1 Concepts   170

5.1.1   Reduction Catalysis Periodic Table   170

5.1.2   Thermodynamics   171

5.1.3 H_2 activation by Metals/Olefin Hydrogenation   172

5.1.4   Hydroformylation: CO Insertion   172

5.1.5   Synthesis Gas Reactions   173

5.1.6   Synthesis Gas: Non-Dissociative CO Adsorption:
MeOH   174

5.1.7   Synthesis Gas: Dissociative CO Adsorption:
Fischer–Tropsch Hydrocarbons   175

5.1.8   Fischer–Tropsch: Schultz–Flory Distribution   176

5.1.9   Hydrodesulfurization   177

5.2 Elements of Catalyst Design   178

5.2.1   Metal Substitution Effect: Fe Versus Co in FT   178

5.2.2   Metals/Ligand Effect: Hydroformylation   180

5.2.3   Hydroformylation Ligand Effects   181

5.2.4   Asymmetric Hydrogenation   183

5.2.5   Promoters and Poisons: Noble Metals   184

5.3 Major Industrial Processes   184

    5.3.1   Hydrogenolysis: Hydrocracking and Hydrotreating   184

    5.3.2   Dehydrogenation: Styrene   185

    5.3.3   Hydroformylation/Hydrogenation: Oxo Alcohols   186

    5.3.4   Carbonylation: Methanol to Acetic Acid   187

    5.3.5   Catalytic Reforming   189

    5.3.6   Ammonia Synthesis   190

5.4 Problems   191

5.5 Answers to Problems   195

6. Environmental Catalysis   **199**

6.1 Concepts   200

    6.1.2.   Thermodynamic Considerations   200

    6.1.3   Exhaust Gas Components   201

    6.1.4   Catalytic Conversion of Exhaust Gas   202

    6.1.5   Catalytic Reactions for Removal of Exhaust Pollutants   203

    6.1.6   Catalysts for Removal of Exhaust Pollutants   205

    6.1.7   Catalyst Components and Functions — 3-Way Catalyst   205

    6.1.8   Catalyst Components and Functions — Lean NO_x Catalyst   207

    6.1.9   Catalyst Components and Functions — SCR Catalyst   208

    6.1.10 Catalyst Components and Functions — NO_x Absorber   209

    6.1.11 Catalyst Functions and Components — DOC and DPF   210

    6.1.12 Mechanistic Concepts   211

6.2 Elements of Catalyst Design   212

    6.2.1   Metal Substitution   212

    6.2.2   Promoters and Poisons   212

6.2.3	Catalyst Synthesis	214
6.2.4	Engineering Concepts	214
6.3 Major Catalytic Technologies		215
6.3.1	Diesel Particulate Trap	215
6.3.2	Diesel Oxidation Catalyst	220
6.3.3	Diesel NO_x Reduction — SCR	222
6.3.4	NO_x Trap	226
6.3.5	Gasoline Engines — 3-Way Catalyst	227
6.3.6	Future Technology: Fuels Cells	228
6.4 Problems		230
6.5 Answers to Problems		233
7.	**Catalyst Characterization**	**237**
7.1 Spectroscopic Methods		238
7.1.1	X-ray Diffraction/X-ray Fluorescence	240
7.1.2	X-ray Photoelectron Spectroscopy	242
7.1.3	Extended X-ray Adsorption Fine Structure	244
7.1.4	Fourier Transform Infrared Spectroscopy	245
7.1.5	Laser Raman Spectroscopy	246
7.1.6	Nuclear Magnetic Resonance Spectroscopy	247
7.1.7	Electron Microscopy	250
7.1.8	Secondary Ion Mass Spectrometry	251
7.2 Adsorption Methods		252
7.2.1	BET Surface Area	253
7.2.2	Porosimetry — Pore Volume and Pore Size Distribution	254
7.2.3	Thermogravimetric Analysis/Differential Scanning Calorimetry	254
7.3 Problems		256
7.4 Answers to Problems		257
References		**259**
Index		263

Foreword

Why should we study catalysis?

- The vast majority of high-volume chemicals are produced using catalysis
- Catalysis is the most cost-effective form of chemistry: the highest cost materials are recycled
- Catalysis is the key to solving many of the world's environmental problems

Topics

- Introduction to Industrial Catalysis
- Acid Catalysis
- Oxidation Catalysis
- Polymerization Catalysis
- Reduction Catalysis/ Hydrogenation
- Environmental Catalysis
- Catalyst Characterization

Within the context of chemistry as a business, catalysis provides the means by which the vast majority of chemicals (on a volume basis) are produced. This can be understood based on the key features of a catalytic process — the efficient conversion of the lowest-cost raw materials to value-added products by processes in which the highest-cost components that effect the reaction are regenerated. In the chemical synthesis laboratory, where the formation of the desired products in the highest purity possible is more important than the cost of its production, these materials are called "reagents," which are consumed in the process and discarded. Every chemist has used such reagents at some point, for example, potassium permanganate for oxidation or lithium aluminum hydride for reduction. Of course, for very high-value products, this strategy also makes commercial sense.

But even in those cases, there are growing environmental and economic driving forces for the use of catalysis as a means to reduce or prevent pollution as opposed to end-of-pipe waste treatment. Pollution that is not produced in the first place makes more cost-effective use of raw materials and reduces environmental impact of chemical operations.

Catalysis has also been a key technology for solving the world's growing emissions limitations, especially for passenger cars and heavy-duty diesel engines for which catalysts are the heart of commercialized emissions reduction devices. Future catalytic scientists and engineers will play a key role in meeting future emission and fuel economy requirements.

Catalysis is truly an interdisciplinary technology, the practice of which requires knowledge of many disciplines, among them, chemistry, chemical engineering and material science. Each of these studies, in their own right, defines a distinct and coherent body of knowledge and each contributes an important perspective on catalysis. And yet, catalysis remains, first and foremost, a chemical phenomenon, and its successful application starts with an understanding of the underlying chemistry. This text attempts to provide a chemical basis for that understanding as a primer for future industrial chemists.

The text is composed of seven chapters: the introduction; four chapters dealing with the major catalytic reaction types of commercial importance: acid catalysis, oxidation, polymerization, reduction/hydrogenation; a chapter devoted to catalytic processes for environmental applications, mainly exhaust gas treatment; and one of methods for analytical characterization.

Chapter 1 (Introduction) is an overview of the history and basic concepts of catalysis and commercial catalytic processes. Chapters 2–6 are presented in three sections: basic Concepts, which covers the fundamental chemical transformations and mechanistic principles, elements of Catalyst Design, which discusses the catalyst parameters important for design of industrial catalysts, and major industrial processes: reaction chemistry and mechanism and catalyst characterizations. Chapter 7 presents the principles and applications of several key analytical methods for catalyst characterization.

In addition to the core areas of study, there are also embedded in the chapters "Toolkit" Topics, which will help the student to master skills associated with the practice of industrial catalysis.

These include:

(1) The representation of catalytic mechanisms using the Tolman formalism,
(2) Methods for the preparation of industrial catalysts, especially heterogeneous catalysts,
(3) Elements of catalyst process engineering, including measurement of process and catalyst efficiency and catalyst design,
(4) Application of methods of analysis for a given catalyst and reaction situation,
(5) Analysis of studies in catalysis from the scientific literature.

Toolkit Topic discussions will include practical examples for students to use in the study and mastery of these skills.

Catalysis "Toolkit" Topics

- Representing catalytic mechanisms
- Methods of catalyst preparation
- Catalysis process
- Engineering concepts
- Catalyst characterization/analysis
- Catalysis literature

1

An Introduction
to Industrial Catalysis

While catalysis is a multidisciplinary science, it is first and foremost a chemical phenomenon, and it is the intent of this text to emphasize its chemical aspects. At the heart of catalytic chemistry is the mechanism by which reactants interact with catalyst, catalytic intermediates are interconnected and products are formed. The foundations of mechanism covered in this chapter will be repeatedly used in subsequent discussions of the various catalytic reaction types.

Another important distinction in catalytic science and technology is that of homogeneous versus heterogeneous catalysis. It is not just a distinction based on the number of phases present in the reaction, but one which also encompasses fundamentally different chemistries and process engineering, with distinct advantages and disadvantages. Finally, in this chapter there will be a summary of the major industrial processes and reaction types, covered in greater detail in each of the subsequent chapters.

1.1 Historical Background

As with any new endeavor, it is useful to begin our studies with a historical perspective, a set of definitions, a few basic concepts and some examples which will provide the basis on which the rest of the course will be built.

A basic understanding of kinetics versus thermodynamics, and the concepts of catalyst activity, selectivity and life will be essential to the discussions in all of the remaining chapters.

Table 1.1.1: Historical Background Early Catalysis Examples[1,2]

BC
• Fermentation of sugars to alcohol

Middle Ages
• Nitric acid catalyzed conversion of sulfuric acid from S, H_2O and air

1812
• Conversion of starch to sugar by dilute acetic acid (Kirchhoff)

1818
• Decomposition of H_2O_2 by Ag, Au, Pt on Mn (Thenard)

1820
• Oxidation of alcohol to acetic acid over Pt (Dobereiner/Davy)

1822
• Combination of H_2 and O_2 by finely divided Pt (Dobereiner)

1834
• Formation of ether on contact with sulfuric acid (Mitsherlich)

Refs. 1, 2

1.1.1 *Early Catalysis Examples*

We begin with a discussion of the history of the discovery, application and understanding of the phenomena we now know as "catalysis." (Table 1.1.1) One of the oldest catalytic processes to be discovered involves not a chemical catalyst, but a biological one. Fermentation, the process by which alcohol is produced from sugars, uses nature's catalysts, the enzymes which are found in yeast, and dates back to many years before the birth of Christ. Only much later in the 19th century did many acquire the knack for effecting catalysis by chemical means. Early examples of chemical catalysis include the formation of sulfuric acid from sulfur, the decomposition of H_2O_2 to H_2O and O_2 by Ag, Au, Pt or Mn (Thenard, 1818) and the oxidation of alcohol to acetic acid over Pt (Dobereiner/Davy, 1820). The catalytic combustion of hydrogen over platinum, and the sulfuric-acid catalyzed dehydration of alcohol to form ether are also examples of catalytic reactions discovered in the 19th century.

1.1.2 *Highlights of Catalysis Technology*

The insight to recognize that there was something unique about these chemical reactions is remarkable, especially considering that the theory of chemical equilibria and "chemisorption" (i.e., the chemical adsorption of reactants onto a catalyst surface) would not be developed for another 40–70 years by van't Hoff and Langmuir, respectively (Table 1.1.2). The development of the synthesis of ammonia from N_2 and H_2 represents the first real commercial catalytic process, which essentially began the large-scale use of N-based fertilizer, upon which modern agriculture is based, and provided feedstock for commercial explosives and dye production. It was the first catalyst which was designed according to first principles of thermodynamics, the understanding of the effects of "promoters" (elements that enhance a catalytic performance) and "poisons" (those which degrade it), and the rational scale-up using critical engineering parameters. The production of synthetic fuels via Fischer–Tropsch catalysis had its origins in the 1920s and was the first practical demonstration of the concepts of chemisorption. The rest of the 20th century saw the introduction of the fuels for the internal combustion engine, petrochemicals and polymers, most of which directly result from the development of commercial catalytic processes. Catalytic cracking, alkylation and reforming of hydrocarbons, first discovered in the 1930s still provide the basis for modern production of fuels and basic chemical raw materials. The foundation for the entire polymers and plastics industry also found its roots in Ziegler–Natta and metal oxide-based polymerization and metathesis catalysts, were also discovered during this decade.

After World War II, major innovations continued, with the discovery that propylene

Table 1.1.2: Highlights of Catalysis Technology[1,2]

1890
- Theory of chemical equilibria (van't Hoff)

1910
- Systematic study of ammonia synthesis (Haber–Bosch)
- Application of thermodynamics (NH_3-decomposition)
- Effect of promoters and poinsons (on Fe catalyst)
- Scale-up issues: High pressure technology

1920s
- Principles of chemisorption — CO oxidation (Langmuir)
- Fischer–Tropsch production of hydrocarbon liquids

1930s
- Catalysis of cracking, alkylation, reforming (Ipatieff & Pines)
- Polyethylene — Phillips (Cr-based) and Ziegler ($TiCl_4$)
- Olefin Metathesis — DuPont/ Standard Oil Ins./Phillips

1950s
- Isotactic Polypropylene– $TiCl_3$Catalyst (Zielgler– Natta)
- Hydrodesulfurization (Co/Mo/S) *p*-Xylene to terephtahlic acid (Co, Mn)

(Continued)

Table 1.1.2: (*Continued*)

1960s
- Selective Oxidation Catalysis Using Air/Metal Oxides (Ido, Veach, Grasselli)
- Propylene ammoxidation to acrylonitrile (Bi/Mo/O$_x$ catalysts — Sohio)
- Propylene oxidation to propylene oxide (Mo, Mobil)
- Zeolite cracking catalyis (Mobil)

1970s
- Shell Higher Olefin Process — α-Olefins from ethylene (Ni/Mo catalyst) Shape Selective Catalysis — ZSM-5 zeolite (Wiesz)
- Acetic acid from methanol — carbonylation, Rh (Monsanto) Rh-based hydroformylation to oxo-alcohols (UCC/Celanese)
- Zeolite hydroisomerization (Shell/UCC)

1980s
- Gasoline from methanol (Mobil)
- Biocatalysis/absymes and catalytic antibodies (Lerner, Schultz)
- Vinyl acetate from ethylene and acetic acid (Pd)
- MTBE from methanol and isobutylene on Amberlyst™ 15

(*Continued*)

could be stereoselectively polymerized to form the isotactic material, which is the basis for modern engineering plastics. Also discovered in the 1950s were hydrodsulfurization catalysis, which is the means by which clean fuels (i.e., low S) are produced, and an inexpensive, selective oxidation route to terephthalic acid, an important monomer in the plastics and coatings industries.

Further innovations based on selective oxidation in the 1960s produced acrylonitrile and propylene oxide, new key monomers for development of nitrile resins and polyethers. Many of today's basic engineered plastics, coatings and performance chemicals are based on these materials.

A key discovery for refinery operations (in the 1960s and 1970s) was the application of "crystalline" forms (i.e., those with long range solid state structural order) of silica-alumina clays called "zeolites" as active and long-lived, hydrocarbon cracking catalysts. Equally important was the discovery that these materials could be designed with molecular cavities and that selective production of desired isomers could be effected based on the size and shape of these cavities.

The 1970s also saw the application of catalysis for use of CO as an organic building block for production of basic fuels and chemical raw materials like methanol and oxo-alcohols (used in many performance and additive chemistries).

Besides further advancements in fuels (methanol and MTBE) and monomers, the 1980s also saw the first concerted efforts to try to mimic biological systems by synthetic design of catalysts based on modern methods of molecular biology (Table 1.1.2). Foremost among these efforts was the use of monoclonal antibodies specifically produced by "antigens" designed to mimic a catalytic transition state, the so-called "abzymes" (antibody enzymes). While practical applications of these systems are still being developed, this technology demonstrates an important marriage of chemistry and biology and is the ultimate demonstration that we now have the ability to produce a catalyst by design for virtually any process for which a molecule that approximates the structure of a transition state can be synthesized.

The discovery of single site/narrow product distribution (versus conventional multisite/ broad product distribution) catalysts have provided further control over desired polymer properties. Multicomponent three-way (NO_x-, hydrocarbon- and CO-reducing) catalysts developed in the 1990s provide the technology for virtually all of today's gasoline–powered vehicle exhaust catalysts.

The 21st century will continue to see advances in catalytic science and technology and its application to critical problems and new opportunities. Foremost among those are related to environmental issues, including emissions reduction, contaminant management and pollution prevention. Specific examples include the replacement of stoichiometric reagents with catalysts, replacement of traditional waste-forming homogeneous catalysts in batch processes with recyclable and

Table 1.1.2: (*Continued*)

1990s
- Three-way emission control catalysts (Pt/Rh/La on Al_2O_3)
- Single-site polyolefin catalysts (Brookhart)

2000s
- "Nano" catalysts
- Diesel exhaust emissions catalysts

2010+
- Improved emissions reduction
- Contaminant Management
- Pollution Prevention

Refs. 1, 2

Table 1.2: Catalyst Definition[1-3]

Berzelius (1835)
- Catalysts awake sleeping affinities by substances without temperature increase

Ref. 3

Leibig (1835)
- Catalysts function by overcoming an 'inertia' that prevents unstable substances from taking part in chemical changes thus accelerating processes which normally occur very slowly in its absence

Ref. 3

Faraday (1834)
- Catalysis requires simultaneous surface absorption of reactants

Ref. 1

Ostwald (1853)
- Catalysts change neither reaction energetics nor its reverse equilibrium

Ref. 1

Modern
- A catalyst accelerates a chemical reaction rate but is not a stoichiometric reactant

Ref. 2

regenerable heterogeneous ones in continuous processes. The familiarity of students with the basic chemical concepts of industrial catalysts will be essential if catalysis is to contribute solutions to these important global problems.

1.2 Catalysis Definitions (Table 1.2)

The discussions to this point regarding the history of catalyst discovery and application have all been made without actually defining what a catalyst is. These discussions are possible because, even in the absence of a rigorous definition, the reader will have a general concept of what a catalyst is in a general sense, even though it may not be absolutely scientifically correct. The word has even found its way into common language, especially in connotations which personify the concept. For example, people who are good at facilitating communication are said to "catalyze" discussions, and people who make things happen can be called a "catalyst." In modern language, being a "catalyst" is generally thought of as a positive attribute.

As the following pages will show, so, too, it is with a reaction that is "catalytic."

As we have seen, examples of catalytic reactions date from before the time of Christ. But in fact, it was not until 1835 that Jacob Berzelius, the Swedish chemist generally regarded as the father of chemical catalysis, coined the phrase "catalytic power," which he defined as the ability to "awaken sleeping affinities" that substances have for one another. About the same time, Leibig defined catalysis as a kinetic phenomenon in which a process is accelerated that otherwise takes place very slowly, and Faraday recognized the need for reactants to be adsorbed simultaneously on a heterogeneous catalyst surace. Ostwald, in 1835, also recognized that catalysts do not effect the state equilibria. These early definitions of catalysis incorporate the essential features of reaction acceleration (kinetics), without affecting equilibria (thermodynamics) (Table 1.2).

As was recognized by 19th century scientists from their definitions stated above, catalysis is a *kinetic* or chemical reaction *rate*-related phenomenon. To understand how a catalyst works, one needs to separate kinetic (rate) issues from thermodynamic (equilibrium) ones.

1.3 Kinetics and Catalysis

First, let us deal with kinetics. Consider the case of the conversion of Reactant A to Product B (Fig. 1.3.1). When A is heated, it is converted to B by path a with an activation energy $= Ea_{uncat}$, which corresponds to the energy difference between the ground state A and the transition state A* (the highest point of path a). This activation barrier defines a rate constant $k = e^{-Ea/RT}$. Thus, the lower Ea, the faster the reaction rate. When a catalyst is added, another, less energy-demanding path becomes available, with activation energy Ea_{cat}, which is lower than Ea_{unct}, resulting in a faster reaction rate. The magnitude of the rate enhancements can be dramatic, for example, the hydrolysis of urea catalyzed by urease enzyme, where the (urease) catalyzed reaction rate is 101^4 times as fast as the uncatayzed rate (Eq. 1.3.1).

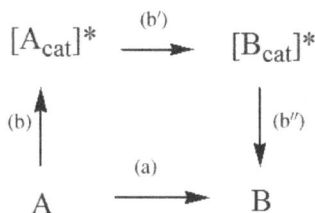

$$H_2NC(O)NH_2 + H_2O \xrightarrow{\text{Urease}} 2NH_3 + CO_2 \quad K_{cat}/K_{uncat} = 10^{14}. \quad (1.3.1)$$

Fig. 1.3.1   Energy Diagram for Catalyzed and Uncatalyzed Conversion of A to B

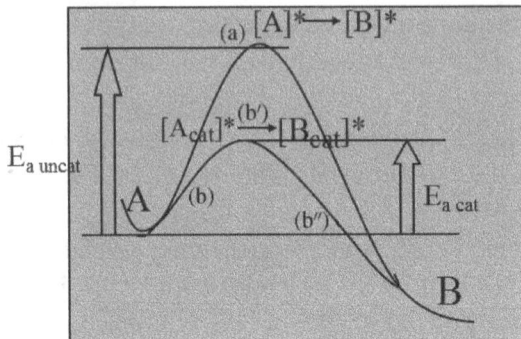

Fig. 1.3.2 **(a) Uncatalyzed Route (b) Catalyst Activation of Reactant A (b′) Conversion of Activated A to Catalyst-Adsorbed Product B**

A catalyst "activates" Reactant A by formation of a Reactant A-catalyst complex $[A_{cat}]^*$ (Fig. 1.3.2). The formation of $[A_{cat}]^*$ occurs by a process called "chemisorption" (chemical adsorption) of A onto the catalyst (path b). $[A_{cat}]^*$ is then converted to a Product B-catalyst complex $[B_{cat}]^*$, by a process whose highest energy state (the transition state) is lower than that of the uncatalyzed path a. The Product B-catalyst complex $[B_{cat}]^*$ rapidly releases or desorbs product (path b″) and the free catalyst is then available to catalyze the conversion of more A to B. While this simplified case does not cover the details of all the transition states and intermediates and their corresponding energetics, it does provide a basic working knowledge of how most catalysts operate. Notice the relative energy levels of ground state A and B are not changed by the presence of the catalyst.

1.4 Thermodynamics

This leads now to the thermodynamic considerations. We have just seen in Section 1.3 how a catalyst affects the *rate* at which a reaction occurs, but not the ground state *energy levels* of reactants and products. Given the same reactants and products, reaction *rate* is accelerated, but not its *thermodynamic equilibrium*. In other words, if a reaction is not *thermodynamically* favored (i.e., ΔG is not <0) without a catalyst, the addition of a catalyst will not make it become so. A catalyst cannot accelerate a reaction that would not already be expected to occur if the system were allowed to come to chemical thermodynamic equilibrium.

For example, the hydrolysis of esters (Eq. 1.4.1, Table 1.4.1) is nearly a energy-neutral process ($K_{eq} \sim 1$). If one uses an equal molar concentration of ester and acid (Table 1.4.2, Case I), one can only expect to achieve an equal-molar concentration of acid and alcohol, i.e., a 50% conversion to acid (by solving Eq. 1.4.2).

The addition of a catalyst will not change the equilibrium constant. However, if one uses a 100-fold excess of water (for example, by using it as the solvent for the reaction as in Case II, Table 1.4.2), the reaction can be driven to 99% completion by mass action (driven equilibrium). In Case I, one would expect a catalyst to accelerate the rate at which ester is converted to acid, but it will not increase its overall conversion. To do that, one would need to use an excess of water or otherwise drive the equilibrium of reaction 1.4.1 to the right. In fact, to achieve both rapid reaction and high conversion, a catalyst (acid) *and* excess water are required to achieve both low reaction times and high ester conversion.

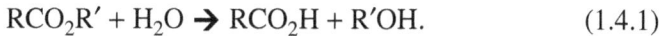

$$RCO_2R' + H_2O \rightarrow RCO_2H + R'OH. \tag{1.4.1}$$

Table 1.4.1: Ester Hydrolysis

Case 1:	1	1	0	0
	$1-x$	$1-x$	x	x
Case 2:	1	100	0	0
	$1-x$	$100-x$	x	x

If $K_{eq} = 1$, then

$$[RCO_2H][R'OH]/[RCO_2R'][H_2O] = 1. \tag{1.4.2}$$

Table 1.4.2: Ester Hydrolysis Cases

Case I: $[H_2O]in = [RCO_2R']in$	Case II: $[H_2O]in = 100[RCO_2R']in$
$x^2/(1-x)^2 = 1$	$x^2/(1-x)(100-x) = 1$
$x_2 = 1-2x+x^2$	$x^2 = 1-101x+x^2$
$2x = 1$	$101x = 1$
$x = 50\%$; $1-x$ (conv.)= 50%	$x = 1\%$; $1-x$ (conv.) = 99%

Non-Oxidative Conditions: C Formation

$$(CH_2)n \rightarrow nC \text{ (coke)} + nH, \tag{1.4.3.1}$$

Oxidative Conditions: Total Combustion

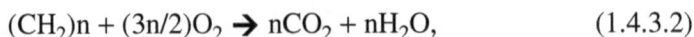

$$(CH_2)n + (3n/2)O_2 \rightarrow nCO_2 + nH_2O, \tag{1.4.3.2}$$

Hydrogenation: Hydrogenolysis/Methane

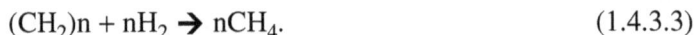

$$(CH_2)n + nH_2 \rightarrow nCH_4. \tag{1.4.3.3}$$

1.4.3 *Thermodynamic Limitation*

Since a catalyst can have no effect on thermodynamics, this will be a limiting factor in the successful application of catalysis to a given reaction of interest. In other words, for a given set of reactants and reaction conditions, there will typically be an ultimate equilibrium state that is undesired and will compete with the desired selective catalyst reaction path. These unselective but thermodynamically–favored products will be discussed in subsequent chapters in the context of each of the major reaction types. However, some generalizations can be made here to acquaint the reader with these limitations as shown in Eqs. 1.4.3.1–1.4.3.3.

Under non-oxidative/non-reductive conditions, the ultimate thermodynamically–favored process is loss of hydrogen to form carbon, otherwise known as "coke" (Eq. 1.4.3.2). This process is known as pyrolysis of hydrocarbons and is a major unselective competing reaction when hydrocarbons are catalytically converted in the absence of oxygen (air). Coke formation not only consumes starting material and produce undesired products, it can also poison catalytic sites and block catalyst pores, thus cutting off the diffusion pathway for reactants and products.

Under oxidative conditions, i.e., in the presence of oxygen (air), coke is no longer thermodynamically–favored relative to total combusion to CO_2 (Eq. 1.4.3.3). In order to produce the desired selective oxidation products (i.e., with the same number of carbon atoms as the starting material), a catalyst must provide a lower activation energy pathway to these selective oxidation products (e.g., acrolein from propylene) than for CO_2 formation.

By the same logic, conversion of hydrocarbon reactants to selective hydrogenation products requires a catalyst that will result in a faster reaction rate to selective products (e.g., a paraffin) from hydrocarbon reactants (e.g., an olefin) than to methane, the thermodynamically–favored product in the presence of hydrogen (Eq. 1.4.3.3).

1.5 Key Concepts

Among the most important properties of catalytic performance are activity, selectivity and life. *Activity* relates to the overall ability of catalyst to convert reactants *to all* products, or in other words, to increase the reaction rate. *Selectivity* refers to the ability of the catalyst to form desired products, usually measured by the ratio of amount of desired product formed to the total amount of *all* products formed. *Life* is the length of time under reaction conditions that the catalyst remains active. Each of these will be discussed in more detail in the following sections of this chapter.

1.5 Key Concepts

Activity: Overall Reaction Rate
Selectivity: Formation of Desired Product(s)
Life: How Long the Catalyst Remains Active

1.5.1 *Activity: Batch Measures*

Activity refers to that ability of a catalyst to accelerate the overall conversion of raw materials to products. Whether a reaction is run as a batch or a continuous process is an important determinant of how catalyst activity is measured. For batch processes, two common measures of activity are % conversion of a given reactant, and % yield of products, as a fraction of theory.

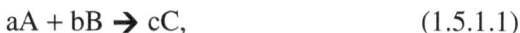

$$aA + bB \rightarrow cC, \qquad (1.5.1.1)$$

Conv. of A =
$$(A_{initial} - A_{final})/A_{initial} \text{ (as a \%),} \quad (1.5.1.2)$$

Conv. to C =
$$aC_{final}/cA_{initial} \text{ (as a \%).} \qquad (1.5.1.3)$$

1.5.1 Activity: Batch Measures

Activity refers to the ability of a catalyst to accelerate the overall rate of conversion of a raw material.

Batch Measures of activity are based on theory:

- % Conversion of a reactant (actual/theory)
- % Yield of a products (actual/theory)

For example, in the reaction of Eq. 1.5.1.1, the % of reactant A remaining equals the difference between the final and initial moles divided by the initial value. The % conversion based on A (i.e., the % of A that has reacted) is 1 (100%) minus the % of reactant A remaining (Eq. 1.5.1.2). The % conversion based on formation of products (in this case Product C), is the moles of C formed divided by the initial moles of A times the stoichiometric ratio a/c (Eq. 1.5.1.3). The batch mode is the simplest and the one most common for small scale reactions and for those run in the laboratory. It also requires very little reaction engineering for the initial investigation of a catalytic reaction, but is less amenable to scale up and requires more labor and operating costs per pound of product produced than continuous processing.

1.5.2 *Activity: Space Velocity*

For a given process, there is a critical production volume at which it becomes most cost-effective to move from batch to continuous processing. In order to represent activity for a continuous process, consider an incremental volume of catalyst over which reactants reside for a given period of time — the residence time — which is the time it takes the reactants to pass through the incremental catalyst volume. The residence time is the volume of the catalyst bed divided by flow rate (Eq. 1.5.2).

Residence Time = vol. of catalyst /
(vol. of reactants/time).   (1.5.2)

While the residence time is analogous to a reaction time in a batch process, it does not,

---

1.5.2 Activity: Space Velocity

For continuous processes, *space velocity* measures absolute conversion (versus a fraction of theoretical) of reactant or product formation per unit time, per unit catalyst used.

Volumetric Space Velocity

- Volume reacted or produced/volume catalyst-time
- GHSV (gaseous hourly space velocity, in hr^{-1})
- LHSV (liquid hourly space velocity, in hr^{-1})

Weight Space Velocity

mass reacted or produced/ mass catalyst–time

For example, weight per weight per hour (WWH)

by itself, give an absolute measure of the capacity of the catalyst to convert reactants to products. For example, let us say that 100 mL of Catalyst A converts a reactant stream of 100% ethane flowing at 100 mL/min to 100% ethylene. Catalyst B, in order to convert all the ethane, requires a reactant stream of a 10% ethane/90% nitrogen mixture flowing at 100 mL/min, which gives a mixture of 10% ethylene/90% nitrogen. For both catalysts the conversion of ethane to ethylene is 100% and the residence time is 100 mL/(100 mL/min) = 1 min. But Catalyst B must have a lower inherent activity because, for a given amount of catalyst, it can only convert 1/10 the amount of ethane per unit time as can Catalyst A. The parameter which reflects the rate at which a given amount of catalyst can convert reactant(s) to product(s) per unit time is called "space velocity", which is the conversion of reactants to products per unit time, per unit of catalyst. Space velocities can be expressed on a volumetric (gas or liquid) or a weight basis and are usually expressed in units of inverse time.

1.5.3 *Space Velocity Examples*

As for batch processes, the activity of a catalyst for continuous processes (in this case as space velocities) can be expressed in terms of either reactants converted or products formed, i.e., the % conversion of reactant or % yield of products per unit catalyst per time. Typical space velocities for three types of catalytic applications, namely chemicals, fuels and exhaust gas conversion, are shown in Table 1.5.3. Chemical production (per unit of catalyst) is about an order of magnitude lower than fuels, which reflects the production capacity of a typical chemical plant versus a refinery.

Because of the rapid gaseous flow rates and the limited space under a vehicle, exhaust gas catalysts require yet another order of magnitude of

Table 1.5.3:   Space Velocity Examples

Process	WWH* or LHSV#, hr^{-1}	GHSV^, hr^{-1}
Chemical Production	0.01–1	10–1,000
Fuels — Refinery	1–10	1,000–10,000
Exhaust Conversion	10–100	10,000–100,000

*Weight /weight-hr.
#Liquid hourly space velocity (vol/vol-hr).
^Gaseous hourly space velocity (vol/vol-hr).

activity. When expressed as gaseous hourly space velocity, the absolute space velocity values are about three orders of magnitude higher than weight or condensed phase volumetric. This reflects the expansion of a liquid when vaporized.

For example, a typical continuous catalytic reactor at a chemical plant might contain one million pounds of catalyst, which has a capacity of one million lbs product per day. This would require a space velocity of:

$$\begin{aligned} \text{WWH} &= 1{,}000{,}000 \text{ lbs product/} \\ &\quad 1{,}000{,}000 \text{ lbs catalyst} - \text{day} \qquad\qquad (1.5.3) \\ &= 1 \text{ day}^{-1} \ (1 \text{ day/24hr}) \\ &= 0.04 \text{ hr}^{-1}. \end{aligned}$$

The same amount of catalyst might be required to convert 100,000 barrels (30,000,000 lbs) of oil to gasoline and diesel fuel in a refinery, which would require a space velocity of $30 \times 0.04 \text{ hr}^{-1}$ or 1.2 hr^{-1}. Exhaust gas catalysts typically occupy about 5 L and need to handle exhaust gas flows of 250,000L/hr, resulting in a space velocity requirement (GHSV) of $250{,}000/5 = 50{,}000 \text{ hr}^{-1}$.

1.5.4 *Activity: Turnover Rate*

Percent conversion, percent yield and space velocities give useful information on the activity of a catalyst per unit volume or weight, which is essential for sizing a reactor to fit a desired capacity. However, these measures give no real information concerning the efficiency of the catalyst on a molecular scale. This type of information is required for one to understand how well a catalyst is performing chemically (versus theory), and how to improve catalyst efficiency. It is also useful in comparing performance of various catalysts of different physical properties and structures and different reaction types.

$$\begin{aligned} \text{Turnover Rate} &= (\text{moles product})/[(\text{moles sites})(\text{time})] \\ &= (\text{g Reactant/MW Reactant})/[(\text{g cat})(\text{sites/g})(\text{hr})] \\ &= (\text{g Reactant/g cat-hr})/[(\text{MW Reactant})(\text{sites/g})] \\ &= \text{WWH}/[(\text{MW})(\text{SD})]. \qquad\qquad\qquad\qquad (1.5.4) \end{aligned}$$

One such measure is called "turnover number" or "turnover rate". Turnover rate is the number of catalytic cycles per active site of catalyst per time, or the number of moles of reactant converted (or product formed) per mole of catalytic active sites. This gives a measure of activity on the molecular level, and therefore requires some measure of the number of molecular entities on the catalyst which perform the catalytic function. These molecular entities are known as catalytically active "sites" and the number of active sites per unit of catalyst surface area is known as "site density" (SD).

There are several chemical and spectroscopic methods by which site density can be measured, including adsorption of molecules which react by a known mechanism (e.g., by an acid–base reaction) with an active site. Examples of these will be discussed in subsequent chapters as they relate to the various catalysts reaction types, and in greater detail in Chapter 7. A common unit of measure is moles reactant/moles of catalytic site-time, which can be derived from measurement of space velocity, molecular weight and site density (Eq. 1.5.4). TN and turnover rate can also be calculated based on moles product formed/mole of catalytic site-time.

1.5.4 Activity: Turnover Rate

- Turnover rate is the number of catalytic cycles per active site of catalyst per time, or the moles of product formed or reactant converted per mole of catalytic active sites
- Requires some measure of the number of active sites per unit weight of catalyst = SD
- Is a molecule-level measure of catalyst activity

1.5.5 *Turnover Example*

As an example of average turnover rate, consider the polymerization of ethylene in which 1 kg of ethylene is charged with 0.5 g of a Ziegler–Natta (TiCl$_3$/AlR$_2$Cl) catalyst (see

Chapter 4.) A catalyst with this formula has a molecular weight of about 200 g/mole, so if the entire solid was the entire solid was composed of active sites accessible to the ethylene, that would correspond to 1 mole of sites/200g = a site density of 0.005 moles of sites/g. Actually, the number of actual sites would be a fraction of this, but for now, let us use this as an estimate of the upper limit. Using Eq. 1.5.4, one can calculate the WWH (based on 95% conversion), and then the average turnover rate = 54,285 turnovers per hour.

• For example, in the continuous polymerization reaction:

$$n(CH_2 = CH_2) \rightarrow (CH_2 - CH_2)n,$$

1 kg ethylene (873 L) is allowed to flow over 0.5 g (0.5 mL) catalyst ($TiCl_3$/AlR_2Cl) over a 15 minute period to form 0.95 kg of polyethylene (PE) product.

   If the PE catalyst site density = 0.005 mole sites/g, the conversion of Ethylene = $1-(1-0.95)/1 = 95\%$, and the yield of PE = $0.95/1 = 95\%$.

   The flow rate of ethylene in = 873 L/15 min = 58.2L/min, corresponding to a residence time = 0.5mL/(58.2L/min) = 8.6×10^{-6} min = 0.0005 sec. The space velocity of PE (by Eq. 1.5.3) = (0.95)(1000g)/ (0.5g) (0.25hr) = 7600 hr^{-1} and the turnover rate (Eq. 1.5.4) = 7600 hr^{-1}/ [(28g/mole)(0.005mole/g)] = 54,285 hr^{-1} or 54,285 turnovers/hr (moles ethylene reacted/mole of cat. sites per hour).

   TN, (i.e., the number of times a catalyst can go through one catalytic cycle before it is deactivated) is also useful as a test of whether or not a material is actually functioning as a catalyst. A reaction must have a TN > 1 to be considered catalytic. TN can be calculated by the number of moles reactant converted (or products formed) per mole of catalyst (mole of catalytic sites). It is also the average turnover rate for a given reaction time. In the above example:

TN = (0.95)(1000g)/(28g/mole)/[(0.005 mole catalyst/g)(0.5g)]

   = 33.9 moles ethylene/0.0025 moles catalyst

   = 13,571 turnovers.

Or, using the turnover rate times reaction time:

TN = (54,285 turnovers/hr)(0.25 hr) = 13,571 turnovers.

1.5.6 *Selectivity*

Another important concept in catalysis is selectivity. Unlike activity, which is an absolute measure of catalyst reactivity, selectivity is a relative measure of catalytic activity for production of a desired product versus the total amount of reactant consumed or product formed. Selectivity to a given product is, therefore, the yield of that product (also called "conversion" to that product) divided by the conversion of reactant to all products:

$$\text{Selectivity} = \text{Yield/Conversion.} \qquad (1.5.6.1)$$

Yield is, therefore, selectivity times total conversion:

$$\text{Yield} = \text{Selectivity} * \text{Conversion,} \qquad (1.5.6.2)$$

$$\text{For the reaction:} \quad aA + bB \rightarrow cC, \qquad (1.5.6.3)$$

selectivity to Product C is calculated as follows:

Conversion of A = moles A reacted/moles A charged

$$= [(A_{initial} - A_{final})/A_{initial}]. \qquad (1.5.6.2)$$

Conversion to C (Yield) = (moles C formed/moles A charged)(a/c)

$$= [aC_{final}/cA_{initial}]. \qquad (1.5.6.3)$$

Selectivity to C = $[aC_{final}/cA_{initial}]/[(A_{initial} - A_{final})/A_{initial}]$. $\qquad (1.5.6.4)$

So where a = b = c = 1 and $A_{initial}$ is 1 mole A_{final} is 0.1 moles and C_{final} is

0.8 moles, the selectivity to C is:

$$\text{Yield of C/Conversion of A} = 0.8/0.9 = 89\%.$$

1.5.7 Selectivity Example

A specific example of this is the conversion of succinic anhydride to succinimide and succinamide with the initial and final moles of reactants and products shown below:

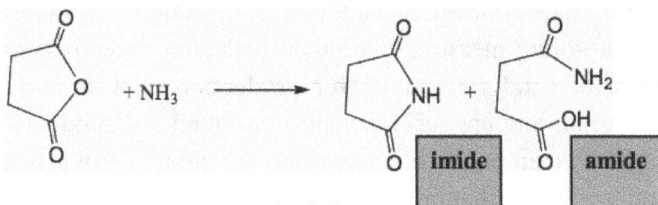

Fig. 1.5.7   Selectivity Example 1

Initial Moles: 1   1   0   0

Final Moles: 0.1   0.1   0.8   0.1

$$\text{Conversion} = 1 - (A_{initial} - A_{final})/A_{initial},$$
$$= 1 - (1 - 0.1)/1 = 0.9 = 90\%,$$

$$\text{Yield of imide} = aC_{final}/cA_{initial},$$
$$= (1)0.8/(1)(1) = 80\%,$$

Selectivity to imide = 80/90 = 88.9%,

Selectivity to amide = 10/90 = 11.1%.

A specific example of this is the conversion of succinic anhydride to succinimide. There are actually two reactions to consider:

$$\text{anhydride} + NH_3 \rightarrow \text{imide},$$
$$\text{anhydride} + NH_3 \rightarrow \text{amide},$$

both of which have reaction coefficients of 1. So the selectivity calculations are straightforward as shown. Notice that the sum of the selectivity to all products is 100%.

1.5.8 *Selectivity Example 2: Formation of Two Products*

If we introduce the formation of a di-imide, there are now three reactions to consider:

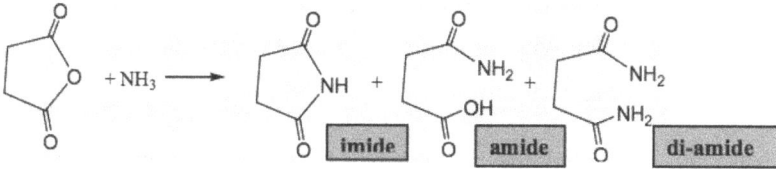

Fig. 1.5.8.1   Selectivity Example 2

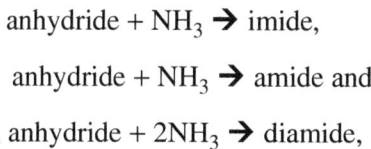

$$\text{anhydride} + NH_3 \rightarrow \text{imide,}$$

$$\text{anhydride} + NH_3 \rightarrow \text{amide and}$$

$$\text{anhydride} + 2NH_3 \rightarrow \text{diamide,}$$

the third of which has a reaction coefficients (c) of 1 for the anhydride and 2 for ammonia. So, the yield and selectivity calculations are now different depending on whether they are based on converted anhydride or converted ammonia.

First, the calculations based on converted anhydride are presented. The yields of imide and amide are the same (80% and 10%, respectively), but now the conversion of anhydride has increased to 95%, resulting in decreased selectivity to these two products and increased selectivity (from zero) to di-amide, so that the total selectivity to all products remains 100%.

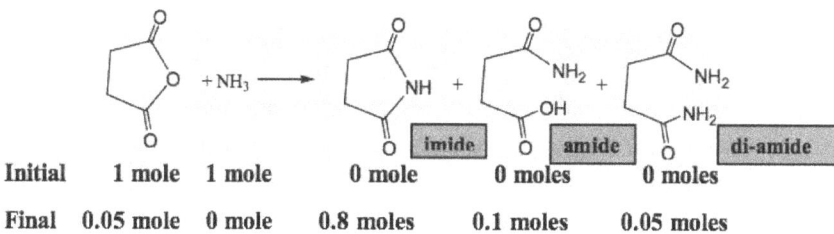

			imide	amide	di-amide
Initial	1 mole	1 mole	0 mole	0 moles	0 moles
Final	0.05 mole	0 mole	0.8 moles	0.1 moles	0.05 moles

Fig. 1.5.8.2   Selectivity Calculations

Based on   **Anhydride Conversion** $= (A_{initial}-A_{final})/A_{initial}$,

converted $= (1-0.05)/1 = 0.95 = \mathbf{95\%}$,

anhydride: **Yield of imide** $= aC_{final}/cA_{initial}$,

                                      $= (1)0.8/(1)(1) = 80\%$,

Selectivity to imide $= 80/95 = 84.2\%$,

Selectivity to amide $= 10/95 = 10.5\%$,

Selectivity to di-amide $= 5/95 = 5.3\%$.

Next are presented the calculations based on converted ammonia. Again, the yields of imide and amide are the same (80% and 10%, respectively) because the total moles of these products formed, the initial moles of ammonia and the reaction coefficients to these products have not changed. But now, the yield of di-imide is doubled because the reactivity coefficient based on ammonia is 2 versus I, based on the anhydride. The selectivities to imide and amide are lower because the conversion of ammonia is 100% (versus 94% for the anhydride) but the selectivity to di-amide is higher because the yield has doubled while the conversion (compared to anhydride) is only 5% higher.

Based on NH_3 Conversion $= 1 - (A_{initial}-A_{final})/A_{initial}$,

converted $= 1- (1-0.0)/1 = 100 = 100\%$,

ammonia: Yield of di-imide $= aC_{final}/cA_{initial}$,

                                      $= (2)(0.05)/(1)(1) = 10\%$,

Selectivity to di-imide $= 10/100 = 10\%$,

Selectivity to imide $= 80/100 = 80\%$,

Selectivity to amide $= 10/100 = 10\%$.

This illustrates that selectivities and yields are dependent on the reactant on which the conversion is based. This is usually the limiting reactant, which in this case would be the anhydride, if ammonia were used in excess. However, if an expensive organic amine (RNH2) were being used to form the substituted imide and amide and di-amide (i.e., substitute an organic R group for one of the hydrogens attached to N in each of these products), succinic anhydride might be used in excess. In this case the reaction yields and selectivities would be based on the amine, which would be calculated as for those based on converted ammonia.

1.5.9 *Catalyst Life*

The last fundamental catalyst property to be considered in this chapter is catalyst life. This corresponds to the length of time that a catalyst remains active. As for activity, life can be measured at the macroscopic or molecular level. In macroscopic/empirical terms, the total units of reactant converted or desired product formed per unit of catalyst. A common unit is weight, which then equates life to (WWH)*(hours of catalyst life) in weight of product that a unit weight of catalyst can produce in its useful lifetime.

On the molecular level, one can express life and the number of catalytic turnovers per active catalyst site during its useful lifetime, which equals the turnover rate times the catalyst life. For the PE example, if one could reuse the catalyst for four of the batches described on 1.5.5, then the life would be (4)×(0.25hr) or 1 hr, and the turnover life would be 54,285 turnovers per catalytic site during a catalyst lifetime.

1.5.9 Catalyst Life

- Corresponds to how long the catalyst remains active
- Catalyst life can be measured in terms of:

— Total units of reactant converted or desired product formed per unit of catalyst used:

  (space vel)(cat life)
  = (wt reactant/wt cat – hr)(hrs activity)
  = wt product that a unit wt of catalyst produces in its useful life

— Turnovers per catalyst lifetime = (Turnover rate) (life)

— For the PE example and a catalyst life of 1 hr, Life=

- (7600 wt/wt-hr)(1 hr) = 7600 wt PE /wt cat or
- (54,285 moles/mole-hr) (1 hr) = 54,285 turnovers/ cat lifetime

1.6 Catalytic Mechanism

The study of catalytic chemistry requires the representation of the pathway or "mechanism" by which chemical bonds are formed and broken, and catalytic intermediates interconverted in the process of converting reactants to products. Chemical reaction mechanisms are useful representations of reaction pathways and the intermediates that are involved in the process. These representations provide insights into how to improve reaction rate (catalyst activity), selectivity to desired products and catalyst life. Strictly

1.6 Catalytic Mechanism
Chemical Reaction Mechanisms:

- Are useful representations of reaction pathways and the intermediates involved
- Provide insights into how to improve a reaction rate, selectivity to desired products and catalyst life
- Cannot be rigorously proven, but incorrect ones can be disproven

speaking, a chemical mechanism cannot be rigorously proven, but incorrect or unlikely ones can be eliminated.

1.6.1 Representing Mechanisms: "Electron Pushing"

- *Electron pushing* is the depiction of how chemical bonds are formed and broken by movement of electron density around and between nuclei of reacting species
- Forms of electrons density include covalent chemical bonds, lone pairs and lone electrons (radical species)
- Electron movement is represented by the use of curved arrows: Double-headed arrows represent motion of two electrons. Single-headed arrows represent motion of one electron
- The origin of electron motion is at the tail end and the destination at the head of the arrow
- Electron pushing is not atom pushing; it does not involve movement of nuclei, only electrons
- The electron pushing arrows are the "word balloons" of the mechanistic chemical cartoon

1.6.1 *Representing Mechanisms: "Electron Pushing"*

A useful formalist in the depiction of chemical mechanism is known as "electron pushing." This is the depiction of how chemical bonds are formed and broken by movement of electron density around and between nuclei of reacting species. Electron density occurs in the form of covalent chemical bonds (molecular species), lone electron pairs (Lewis bases) and lone electrons (free radicals). In the electron pushing convention, electron movement is represented by the use of curved arrows. A double-headed arrow represents motion of two electrons (from a covalent bond or lone pair), while a single-headed arrow represents motion of one electron (e.g., from a free radical or metallic species). The origin of electron motion is at the tail end of the arrow and the destination at its head.

Conventional (i.e., non-nuclear) chemistry is based on redistribution of electron density. The depiction of this in electron pushing does not involve the movement of atomic nuclei. In other words, electron pushing is just that, *not* atom pushing. One way of thinking about this is to consider a mechanism, a kind of "cartoon" of how a chemical reaction might work. In this context, electron pushing notations are the "word balloons" that convey to the reader the "dialogue" between chemical intermediates.

1.6.2 *Representing Mechanisms: "Name Reaction" Examples*

"Common name reaction involving electrons solely in covalent bonds include the so-called "electrocyclic" reactions (electrons in cyclic transition states) such as the Diels–Alder reaction, the ene reaction, and the decarboxylation of malonic acids (shown in Fig. 1.6.2). Examples of reactions involving lone pairs are the keto–enol tautomerism, Cope elimination and SN2 substitution. Notice that in each case, curved arrows in the reacting molecules indicate movement of electron density to the new locations shown in the resulting product structures.

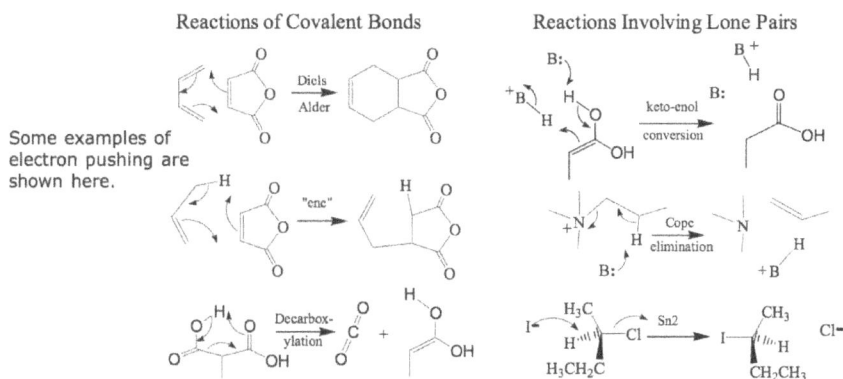

Fig. 1.6.2   Mechanisms of Some Name Reactions

1.6.3 *Elementary Catalytic Reaction Steps*

Some elementary reaction steps which occur repeatedly in catalytic reaction mechanisms are shown in Fig. 1.6.3. *Ligand dissociation* is a means by which catalytic species can create coordination vacancies (□), which serve as active reaction sites for "chemisorption" (chemical adsorption) of reactant (see 1.3). For example, loss of CO from metal carbonyls, especially Co and Rh, is the key step which creates vacant coordination sites for chemisorption of olefins in catalytic hydroformylation (1.6.5 and 5.1.4). Species which contain coordination vacancy are known as being *coordinately unsaturated* and usually have less than an octahedral coordination, i.e., less than 6 ligands, or in other words, less than a coordination number of 6. Such unsaturated species usually also contain less than the

Ligand Dissociation/Association:

Migratory Insertion:

Oxidative Addition:

Reductive Elimination:

β-Elimination:

α-Elimination:

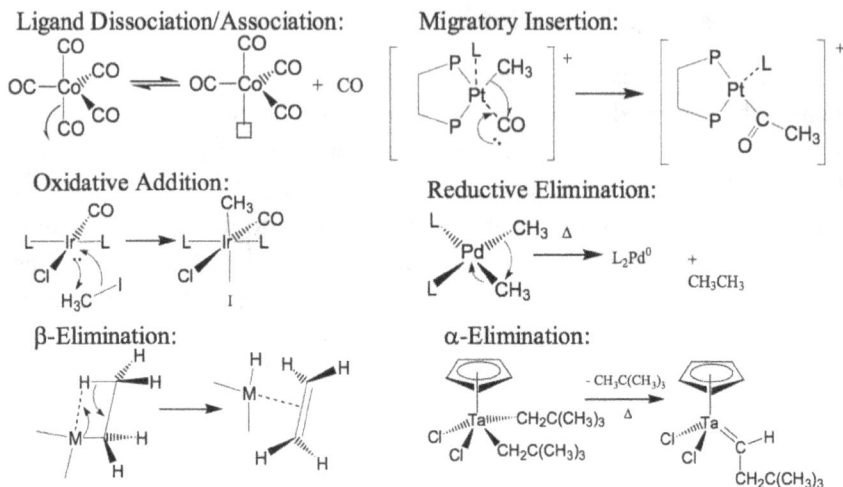

Fig. 1.6.3   Reaction Steps (Adapted from Ref. 1 with permission)

totally filled outer shell of d electrons (usually 16), compared to the corresponding coordinately saturated species with a full complement of ligands and d-electrons (18). Because a filled shell is the most stable electronic state, the coordinately unsaturated species is of higher energy and thus more reactive toward chemisorption of reactant, which re-fills the shell, resulting in a coordinately saturated 18-electron species. Thus, there is a thermodynamic driving force for chemisorption process. The interconversion of such 16- and 18-electron species is common in transition metal catalyzed reactions. There is now good evidence that in many cases, the 16-e metal hydride actually contains a C–M–H "agostic" 3-center 2-e interaction (shown Fig. 1.6.3, "β-elimination") which is actually an 18-e ground state for the β-elimination reaction (Ref. 4).

   Migratory insertion is another common mechanistic step in transition-metal catalyzed reactions which results in interconversion of coordinately unsaturated and saturated species, concurrent with the formation of C–C bonds, such as also occurs in hydroformylation (Figs. 1.6.5 and 5.1.4). *Oxidative addition* and *reductive elimination* steps create intermediates with various degrees of coordinate unsaturation by formal oxidation or reduction of a metal center. In *beta- and alpha-elimination* reactions, organic unsaturation (i.e., where there are < 4 atoms bonded to C) results by formation of carbon-carbon double bonds (C=C) or carbon-metal double bonds (C=M).

1.6.4 *Representing Catalytic Mechanisms: the "Tolman Formalism"*

Catalytic reaction mechanisms are best represented in a form which depicts the activation or chemisorption of reactants by the catalyst, the reaction of catalytic intermediates to form products, and the reformation/reactivation of the catalytically active species. This method of representing catalytic reactions is known as the Tolman formalism. In this method, reaction intermediates are arranged in a circle; one catalytic cycle is represented by one clockwise motion around the circle. The catalytically active species (typically at 12:00) activates reactant and produces product as one moves clockwise around the circle, and ultimately re-forms the active catalyst. All reactions which are catalytic can be depicted in this way.

1.6.5 *Catalytic Mechanism Example: Hydroformylation of Olefins*

An example of a catalytic mechanism in the Tolman Formalism is shown in Fig. 1.6.5 for the hydroformylation of olefins. Notice that reactants come into the circle of catalytic intermediates. In contrast, intermediates and products come out. The activation steps which form the active catalyst, namely:

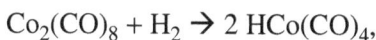

$$Co_2(CO)_8 + H_2 \rightarrow 2\ HCo(CO)_4,$$

$$HCo(CO)_4 \rightarrow HCo(CO)_3 + CO$$

are not actually part of the catalytic cycle, since this one occurs once (theoretically) and the actual active species $HCo(CO)_3$ is reformed in the catalytic cycle. The stochiometric

1.6.4 Representing Catalytic Mechanisms: "Tolman Formalism"

• Involves the reaction/ deactivation and reformation reactivation of catalytic species
• Reaction intermediates are arranged in a circle — one catalytic cycle is once around the circle
• If a reaction cannot be depicted this way, it is not truly catalytic

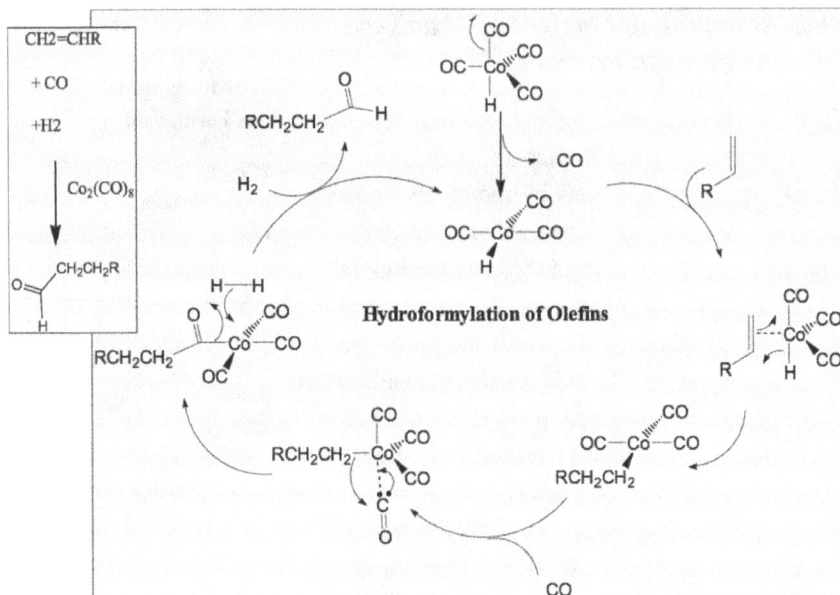

Fig. 1.6.5 Hydroformylation of Olefins (Refs. 5, 6)

reaction can be written by adding up all the reactants which come into the cycle and all the products which form:

$$CH_2 = CHR + CO + H_2 \rightarrow RCH_2–CH_2–CHO.$$

In this case, the active catalyst $HCo(CO)_3$ is a coordinately unsaturated species which forms by loss of CO from $HCO(CO)_4$. This creates a coordination vacancy which serves as the site for chisorption of olefin. After oxidative addition, CO addition, migratory insertion and reductive elimination, the product aldehyde with one more carbon atom is formed. Such aldehydes (e.g., butyraldehyde as shown in Fig. 1.6.5) made by catalytic hydroformylation are the raw materials for "oxo" alcohols, an important high volume chemical intermediate for surfactants and other performance chemicals (5.1.4 and 5.3.3).

1.7 Homogeneous Versus Heterogeneous Catalysis

An important distinction in catalytic processes is whether the catalyst and the reactants are in the same phase (typically the liquid phase), which is

Table 1.7: Homogeneous Versus Heterogeneous Catalysis

	Homogeneous	*Heterogeneous*
Mechanism	Molecular	Surface/Solid
Active Site	Single Metal Center	Multiple Surface Sites
Site Environment	Ligands	Nearest-Neighbor Atoms
Composition	Molecular Formula	Phase Diagram
Active Site	Coordinate Unsaturation	Atomic Vacancies
Activation Step	Coordination	Chemisorption
Product-release	Ligand exchange	Desorption
Diffusion of reactants/products	Solubility/Viscosity	Pore Size

known as *homogeneous* catalysis), or they are in separate phases (typically a solid catalyst and liquid or gaseous reactants), which is known as *heterogeneous* catalysis. While this distinction, based on the number of phases present, is a useful definition, there are much more significant differences in the fundamental chemistry by which these processes occur (Table 1.7).

Mechanistically, homogeneous catalysts are molecular species with an active site, usually composed of a single metal center. The environment around the metal center is composed of ligands whose number and nature are controlled by the thermodynamic stability and reactivity of the corresponding organmetallic species. Activation occurs by formation of coordinately and electronically unsaturated sites via ligand dissociation (for example, loss of CO from a metal carbonyl, as in Fig. 1.6.3), which activate reactants by coordination. Diffusion of reactants to, and products from, the catalytic sites are controlled by their solubilities in and the viscosity of the reaction media.

Heterogeneous catalysts contain multiple sites on the surface of a solid state matrix or bulk phase. The nature and number of nearest neighbor atoms in a catalytic surface site is controlled by surface energetics and oxidation state of the central metal atom, while that of the bulk (for crystalline solids) is controlled by the stabilities of the crystallographic phases at a given stoichiometry — i.e., the phase diagram. Atomic vacancies and/ or coordinately unsaturated (electronically deficient) centers serve as the sites for activation of reactants by chemisorption. Diffusion of reactants

into and desorption/diffusion of products away from the catalyst are controlled by the size and distribution of pores in the catalyst. These fundamental distinctions create advantages and disadvantages for each type of
system (as discussed in 1.7.3 and 1.7.4).

1.7.1 *Homogeneous Catalysis Example*

An example of a homogeneous catalyic process is the epoxidation of olefins by hydroperoxides catalyzed by soluble Mo-oxo complexes of the
formula MoL_2O_2 (OH)(OR) (Fig. 1.7.1). The choice of the ligands (L)
controls the electronic and steric (space-filling) environment around the
active metal center. These complexes are soluble in a number of organic
solvents and are single-metal, molecular catalysts.

Activation occurs by ligand (L) dissociation, which produces a coordinately unsaturated site. Olefin is then activated by coordination to the
resulting vacant site.

The catalytic cycle (Fig. 1.7.1) is composed of intermediates which
interconvert between coordinately unsaturated reactive (5-coordinate trigonal bipyramids) and saturated octahedral (6-coordinate). Product forms by

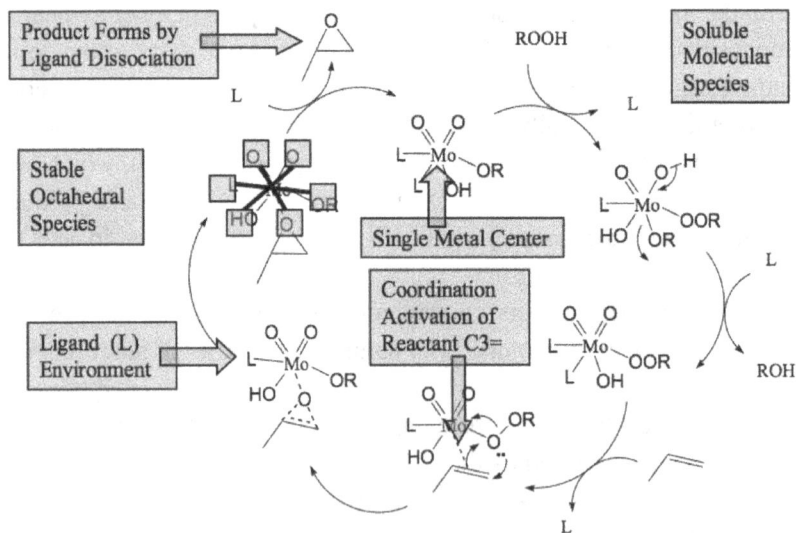

Fig. 1.7.1 Homogeneous Example Epoxidation of Propylene (Ref. 6)

dissociation of epoxide as a ligand from the Mo metal center and dissolution/diffusion into the reaction media. The olefin, in this case propylene ($CH_3CH=CH_2$) and hydroperoxide (ROOH) reactants are thus provided a low energy pathway for conversion to product epoxide and the corresponding alcohol (ROH).

1.7.2 Heterogeneous Catalysis Example

Consider the selective oxidation of propylene to acrolein on solid bismuth molybdate catalysts as an example of heterogeneous catalysis, with the proposed mechanism shown in Fig. 1.7.2. The molar ratio of Bi:Mo in the catalyst is 2:3, which corresponds to a stable oxide bulk phase with an empirical formula $Bi_2Mo_3O_{12}$, known as the *a*-bismuth molybdate. In this case, oxygen atoms (instead of ligands as in the homogeneous case) make up the nearest neighbors of the metal centers. Reactive coordinately unsaturated sites occur as oxygen vacancies (□) and M=O double bonds (versus the vacant coordination sites for homogeneous catalysts.) In heterogeneous catalysts, the stable bulk phase serves as a template for the Bi-Mo cluster

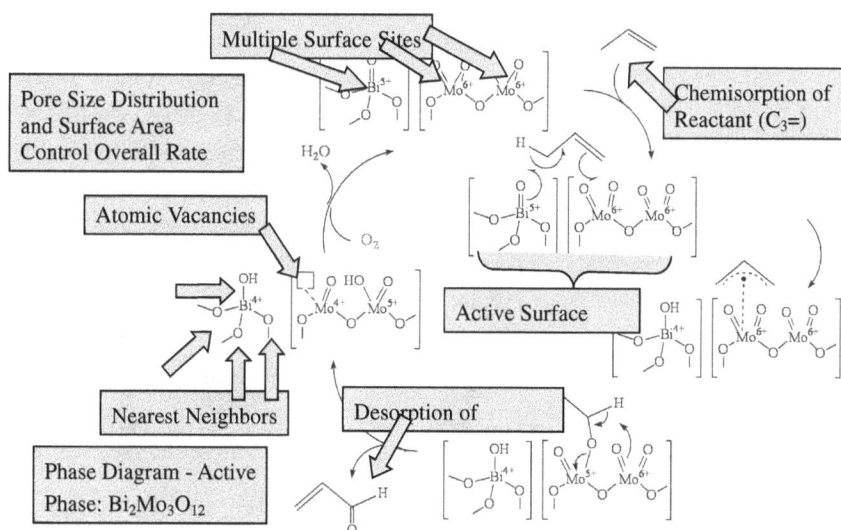

Fig. 1.7.2   Heterogeneous Catalysis Example: Oxidation of Propylene to Acrolein (Adapted from Ref. 7 with permission)

1.7.3 Advantages of Homogeneous Catalysis

- Wide range of molecular compositions available (versus control of composition by phase diagram)
- Rational design of desired catalytic elements into a single molecular species
- A single, homogeneous site may be required to produce a narrow product distribution (e.g., polymers)
- Ease of synthesis by standard methods
- Ease of characterization and testing
- Optimization by classical ligand selection for tailoring electronic and steric effects
- Amenable to *in situ* methods of study

with a Bi:Mo ratio of 1:2, versus the 2:3 ratio in the bulk phase ($Bi_2Mo_3O_{12}$).

Unlike homogeneous single site catalysts, the active sites of a heterogeneous catalyst contain multiple metal centers in several combinations, which are frequently of different compositions than, but stabilized by, the bulk phase. Like the homogeneous case, activation occurs by interaction (in this case by chemisorption versus ligand formation) of reactant (propylene) on unsaturated (Mo^{6+}=O) surface sites and product (acrolein) is formed by desorption from the surface site (versus ligand dissociation). The number of active sites accessible to reactant is controlled by the available surface area of the solid. Diffusion of reactants to the catalyst and product away from the catalyst is controlled by the pore structure of the solid. In contrast, for homogeneous catalysis, available surface area and pore structure are not usually important considerations. As long as the catalyst is soluble in the reaction media, these factors are controlled by diffusion in the media, which, for non-viscous liquids at or above room temperature, are normally much faster than the reaction rate and are therefore not rate-limiting. Because they are insoluble solids, heterogeneous catalysts pose significant challenges to study (especially *in situ*) on the molecular level.

1.7.3 *Advantages of Homogeneous Catalysis*

These fundamental differences between heterogeneous and homogeneous catalysts result in specific advantages for each. Homogeneous catalysts permit the use of a wide range of molecular compositions which may not be within the phase diagram for the analogous solid state system. Homogeneous catalysts can also be rationally designed to incorporate all the desired catalytic elements into a single molecule. For example, metal carbonyls such as those based on Co and Rh can be tuned using phosphine (PR_3) to produce a specific electronic and steric condition around the metal atom with a degree of precision not possible in a heterogeneous system. For certain products, such as polymers, the production of a narrow product distribution requires a narrow distribution of homogeneous sites. For example, solid polymerization catalysts with a broad distribution of active sites will also produce a broad distribution of polymers (i.e., with a broad distribution of polymer chain lengths.) Homogeneous catalysts can be synthesized, characterized and tested by standard methods and can be optimized by tuning of electronic and steric effects by ligand adjustment. And because the active catalysts are discrete molecular species in solution, homogeneous catalyst are well-suited for study and characterization on the molecular level, especially for *in situ* studies, i.e., while the reaction is occurring.

1.7.4 *Advantages of Heterogeneous Catalysis*

For heterogeneous systems, one of the main advantages is the ease of separation of product and catalyst, due to the insolubility of the catalyst in the product. This feature also makes

1.7.4 Advantages of Heterogeneous Catalysis

* Ease of Product/Catalyst Separation
* Recycle and Regeneration Possible
* Thermal and Oxidative Stability
* Low Cost/Large Scale Production
* Activity (surface area) can be increased and cost reduced by supporting:
    — SiO_2 (Silica)
    — TiO_2 (Titania)
    — Zeolites
    — Al_2O_3 (Alumina)
    — Carbon
    — Clays

recycle and regeneration of catalyst more feasible. And because heterogeneous systems are typically composed of thermodynamically stable inorganic solids, they have high thermal and oxidative stability. Their low cost and compatibility with continuous processing also make them ideal for large-scale production. In addition, the available surface area and thus the activity of heterogeneous catalysts can be increased, and the cost reduced, by adsorbing the active catalyst on an inexpensive solid support such as silica, alumina, titania, carbon, clays (amorphous silica–aluminas) and zeolites (crystalline silica–aluminas).

Because of these advantages, heterogeneous systems make up the majority of large-scale commercial chemical production, especially for continuous processing of commodity (lower value) materials, where only very low processing costs per pound of product can be tolerated. However, there are still many commercial examples of homogeneous catalysis, for which it has not been possible to identify a heterogeneous system with the required activity and selectivity to the desired product.

1.8 Major Industrial Catalytic Process Types

• Acid Catalysis
• Oxidation Catalysis
• Polymerization Catalysis
• Reduction Catalysis/ Hydrogenation
• Environmental Catalysis

1.8 Major Industrial Catalytic Process Types

The next five chapters deal with the major industrial catalytic process type. These are:

Acid Catalysis: The activation of reactant by a reaction with a Bronsted (protonic) or a Lewis Acid, which is regenerated in the catalytic cycle. Reactants and products are of the same oxidation state. Major reaction types used extensively in both chemicals and fuels production include hydrolysis, hydration, dehydration, alkylation and isomerization.

Oxidation Catalysis: The acceleration of a selective pathway in the oxidation of an organic reactant, relative to unselective ones (combustion). Selective oxidation is used mainly in the chemicals industry to add value by introduction of polar functionality with desired chemical reactivity and performance properties.

Polymerization Catalysis: The formation of covalent bonds between monomeric units. Polymerization catalysts contain anionic, cationic or radical centers which are the monomer activation sites. Polymers have widespread use across commodity and specialty chemicals markets and account for a significant volume of commercial products derived from catalysis.

Reduction/Hydrogenation Catalysis: The redistribution of hydrogen (H_2) among carbon-containing molecules and with molecular H_2, including the formation and breaking of C–C bonds. Reduction and hydrogenation are the major processes used for refining and upgrading petroleum-based fuels and for production of synthetic fuels.

Environmental Catalysis: The conversion of pollutants in exhaust gas streams into non-pollutants. Major pollutants include NO_x, soot (and other particulates), CO and hydrocarbons. Exhaust gas treatment catalysts have a particularly challenging task of reducing pollutants to parts per million (ppm) levels from a rapidly flowing gas in a very limited space.

The corresponding chapters will cover the basic concepts, elements of catalyst design and the chemistry and mechanism for some of the commercial applications of each process type.

1.8.1 *Industrial Examples*

A summary of some of the specific examples of industrial applications for each of the catalyst process types that we will be covering in detail in subsequent chapters is listed in Table 1.8.1.

Table 1.8.1:   **Examples of Industrial Catalytic Processes**

Acid Catalysis	Catalyst	Type	End Use
Fluid Catalytic Cracking	Zeolites, silica–alumina	Heterogeneous	Gasoline
Paraffin Isomerization	Zeolites, silica–alumina	Heterogeneous	Gasoline
Alkylation	$AlCl_3$, HF, H_2SO_4	Heterogeneous	Gasoline
Methanol-to-Gasoline	Zeolites	Heterogeneous	Gasoline
Ethylbenzene from Ethylene and Benzene	Zeolites	Heterogeneous	Polystyrene
Cumene from Propylene and Benzene	Zeolites	Heterogeneous	Polymers
Oxidation Catalysis			
Ethylene Epoxidation	Ag	Hetereogeneous	Polyethers
Butane to Maleic Anhydride	V/P oxides	Hetereogeneous	Chemical Intermediate
Ethylene to Vinyl Acetate	Pd/Cu	Homogeneous	Polmers
Cyclohexane to Adipic Acid	Co	Homogeneous	Nylon-6
p-Xylene to Terephthalic Acid	V/Ti oxides	Heterogeneous	Polymers/ Plastics
Propylene to Acylonitrile	Bi/Mo/oxides	Hetereogeneous	Polymers
Hydrogenation	**Catalyst**	**Type**	**End Use**
Ammonia Synthesis (Haber–Bosch)	Fe/K	Hetereogeneous	Basic Raw Material
Benzene to Cyclohexane	Ni/Pt	Homogeneous	Nylon-6
Hydrotreating/ Hydrocracking	Co/Mo oxides	Heterogeneous	Fuels
Hydrofomylation	$M(CO)_xLy$ (M=CO or Rh)	Hetereogeneous	Fuels
Ethylbenziene Dehydrogenation to Styrene	Cr_2O_3	Hetereogeneous	Basic Monomer
Polymerization			
Polyethylene and Polypropylene	$TiCl_3/MgCl_2$	Heterogeneous	Thermoplastics
	Cp_2ZrCl_2	Hetereogeneous	

(*Continued*)

Table 1.8.1: *(Continued)*

Hydrogenation	Catalyst	Type	End Use
Ethylene Oligomerization to a-Olefins	Ni/Mo	Hetereogeneous	Chemical Intermediate
Polystyrene	$AlCl_3$, KNH_2 or AIBN	Hetereogeneous	Plastics
Polyisobutylene	$AlCl_3$, BF_3	Hetereogeneous	Lubricants
Environmental			
3-Way Gasoline Catalyst	$Pt/Rh/LaO/CeO/Al_2O_3$	Hetereogeneous	HC, CO and NO_x
Lean NO_x Diesel	Cu/ZSM5 (HT) or Pt/Al_2O_3 (LT)	Hetereogeneous	NO_x
Selective Catalytic Reduction (SCR)	Pt, V_2O_5/Al_2O_3 or Zeolite	Hetereogeneous	NO_x
NO_x Absorber	$Pt/Rh/BaO/Al_2O_3$	Hetereogeneous	NO_x
Diesel Oxidation Catalyst (DOC)	Pt/Al_2O_3	Hetereogeneous	CO, soot
Diesel Particulate Filter (DPF)	Pt, Pd, Rh, or Ru	Hetereogeneous	Soot
	V, Mg, Ca, Sr, Ba, Cu, Ag	Hetereogeneous	Soot

1.9 Problems

1. The following homogeneous catalytic reaction, run in a solution batch process, is used to produce acrylamide, a monomer used in coatings:

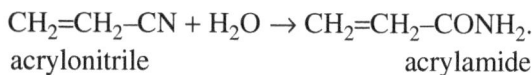

    $$CH_2=CH_2-CN + H_2O \rightarrow CH_2=CH_2-CONH_2.$$
    acrylonitrile   acrylamide

    An undesirable product is acrylic acid formed by the reaction:

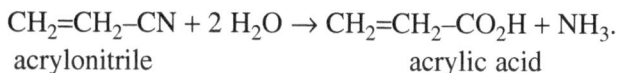

    $$CH_2=CH_2-CN + 2\ H_2O \rightarrow CH_2=CH_2-CO_2H + NH_3.$$
    acrylonitrile   acrylic acid

    Catalyst A, at 80°C reaction temperature and 6 hr reaction time converts 1 mole of acrylontrile to 0.6 mole of acrylamide 0.35 mole acrylic acid with 0.05 moles of unconverted acrylamide.

Catalyst B under the same conditions, converts 1 mole of acrylontrile to 0.3 mole of acrylamide, 0.05 mole acrylic acid with 0.65 moles of unconverted acrylamide.

a. For Catalyst A and for Catalyst B, what is the % conversion of acrylonitrile, the % yield of acylamide and the % selectivity to acrylamide?

b. Which one would you use if you had only 10 moles of acrylonitrile on hand and had to make 6 moles of acrylamide in 8 hrs?

c. Which catalyst would you use if you only needed to make 0.25 mole acrylamide and you could recover unreacted acrylonitrile?

2. The following homogeneous catalytic reaction, run in a continuous liquid-phase process, is used to produce acrylamide, a monomer used in coatings:

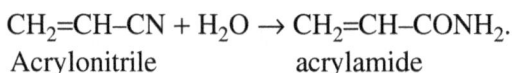

$$CH_2=CH-CN + H_2O \rightarrow CH_2=CH-CONH_2.$$
Acrylonitrile   acrylamide

An undesirable product is acrylic acid formed by the reaction:

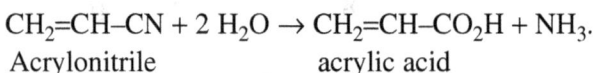

$$CH_2=CH-CN + 2\ H_2O \rightarrow CH_2=CH-CO_2H + NH_3.$$
Acrylonitrile   acrylic acid

The flow rate into the reactor is 5 mL/min, in which is loaded 25 mL of catalyst. The effluent in is a 10% acrylonitrile and 90% water. The flow rate of the effluent out is also 5 mL/min and is 1.1% acrylonitrile, 87.3% water, 10.9% acrylamide and 0.15% acrylic acid. (All %'s are on a Vol. basis)

(a) What is the % conversion of acrylonitrile?

(b) What is the % yield of acrylamide? (assume a density of 1 for all materials)

(c) What is the selectivity to acrylamide?

(d) What is the LHSV of feed through the reactor?

(e) What is the average contact time in the reactor?

3. The selective catalytic heterogeneous oxidation of ethylene is a continuous gas phase reaction run at 425°C over an AgO catalyst:

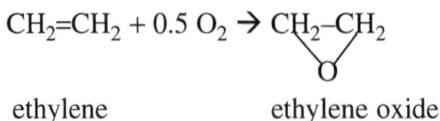

$$CH_2=CH_2 + 0.5\ O_2 \rightarrow CH_2-CH_2$$

ethylene   ethylene oxide

The major competing unwanted reaction is total oxidation to CO_2:

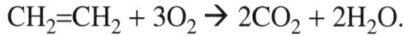

$$CH_2=CH_2 + 3O_2 \rightarrow 2CO_2 + 2H_2O.$$

The reaction is run in the laboratory with 10 mL of catalyst. The flow of reactant gas fed into the reactor (measured at 25°C) is 30 mL/min of the following composition:

21% ethylene
16% O_2
63% N_2

The flow of gases out of the reactor (after water is condensed out) is 26.3 mL/min of the following composition:

2.4% ethylene	19.4% ethylene oxide.
1.8% O_2	4.3% CO_2
72% N_2	

What is the:

(a) % conversion of ethylene?
(b) % yield to ethylene oxide?
(c) % selectivity to ethylene oxide?
(d) GHSV of feed gas through the reactor (Remember the reactor is at 698K, but the gas flow of 30 mL/min was measured at 298K)?
(e) Average residence time?
(f) What is the WWH of ethylene through the reactor? Assume ideal gas behavior ($PV=nRT$, $R=0.0825$ L-atm/mole-deg K) and catalyst density of 1.0)?
(g) At a GHSV of 280 hr^{-1}, what flow rate should be used in a reactor charged with 10 mL catalyst?
(h) At a flow rate of 20 mL/min (298K), what amount of catalyst should be used to give an average contact time of 8.6 sec?
(i) At a flow rate of 30 mL/min (298K), what amount of catalyst should be used to give an average contact time of 4.3 sec? What is the new GHSV?

4. For the following reaction mechanism:

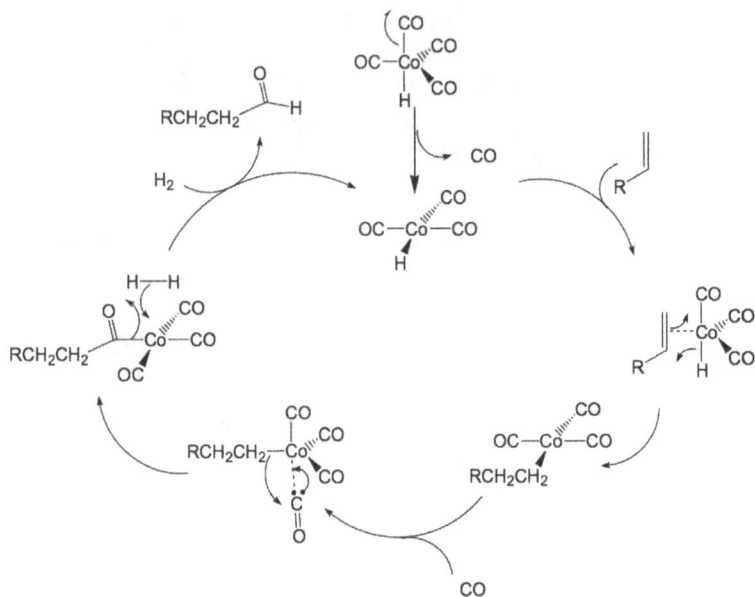

(Ref. 6)

(a) Circle the coordinately unsaturated metal species.
(b) Draw a square around the coordinately saturated metal species.
(c) Put an X over the arrow of the ligand dissociation steps.
(d) Draw a double-headed arrow between the olefin-hydride and metal alkyl.
(e) Write an "MI" over the arrow of the migratory insertion step.
(f) Write a balanced equation for the overall reaction mechanism (from Ref. 6).
(g) Is $HCo(CO)_4$ the catalyst? Why or why not?
(h) Do you think this is a homogeneous or a heterogeneous catalytic reaction? Why or why not?

1.10 Answers to Problems

1. a. Cat A: Conv = $(1-0.05)/1$ = 95%; yield = $0.6/1$ = 60%; selectivity = $60/95$ = 63%.
        Cat B: $(1-0.65)/1$ = 35%; yield = $0.3/1$ = 30%; selectivity = $30/35$ = 85.7%

b. Cat A.
Cat A could make:
(0.6 moles acrylamide/mole acylaonitrile-6hr)*10 moles acrylonitrile * 8 hr = 8 moles acrylamide
Cat B could only make:
(0.3 moles acrylamide /mole acrylonitrile-6 hr)*10 moles acrylnitrile* 8 hr = 4 moles acrylamide
c. Cat. B. Cat A is less selective and would waste more acrylonitrile than Cat B.

2. a. (5 mL/min)(10%) = 0.5 mL/min in acryloitrile
(5 mL/min)(1.1%) = 0.055 mL/min out acrylonitrile
(0.5–0.055)/0.5 = 89% conversion
b. What is the % yield of acrylamide? (Assume a density of 1 g/ mL for all materials.)
0.055 mL/min = 0.055 g/min/53 g/mole = 0.00104 mole acrylonitrile/min in
(0.5 mL/min)(0.109) = 0.0545 mL/min = 0.0545 g/min out acrylamide out
0.0545 g/min/(71 g/mole) =0.000768 mol acrylonitrile/min out
0.000768 mol acrylamide out /0.00104 mole acrylonitrile in = 73.8% yield
c. Selectivity = yield/conversion = 73.8/89 = 81.1% yield
d. LHSV of feed through the reactor = (5 mL/min)/25 mL = 0.2 min^{-1} (60 min/hr) = 12 hr^{-1}
e. 1/12 hr^{-1} = 0.00833 hr *(60 min/hr) = 5 min, or more simply: volume cat/flow rate = 25 mL/5 mL/min = 5 min

3. a. % conversion of ethylene
Average flow of ethylene in = (0.21)(30 mL/min) = 6.3 mL/min
Average flow of ethylene out = (0.024)(26.3 mL/min) = 0.63 mL/min
(6.3–0.63)/6.3 = 90% conv.
b. % yield to ethylene oxide
0.194 × 26.3 mL/min ethylene oxide out = 5.1 mL/min
5.1/6.3 = 80.9% yield
c. % selectivity to ethylene oxide
Selectivity = yield/conv = 80.9/90 = 90%
d. GHSV of feed gas through the reactor (Remember the reactor is at 698K, but the gas flow of 30 mL/min was measured at 298K.)

(30 mL/min) at 25°C (298K)

Flow of gas in reactor (698K) = (698/298)*30 mL/min = 70.3 mL/min

$GHSV = (70.3 \text{ mL/min})/10\text{mL} = 7.0 \text{ min}^{-1}$ (60 min/hr) = 420 hr^{-1}

e. Average residence time

$= 1/GHSV = 1/7 \text{ min}^{-1} = 0.143$ min = 8.6 sec

f. WWH of ethylene through the reactor? Assume ideal gas ($R = 0.0825$ L-atm/mole-K) and catalyst density of 1.0.

$PV = nRT$

$n = PV/RT$

$n = (1 \text{ atm})(0.0063 \text{ L/min})/[(0.0825 \text{ L-atm/mole-°})(273°C)]$

$n = 0.00028$ mole/min ethylene into the reactor

(0.00028 mole/min)(28 g/mole) = 0.0078 g/min

WWH = (0.0078 g/min)(60 min/hr)/10 g = 0.47 hr^{-1}

g. At a GHSV of 280 hr^{-1}, flow rate (setting at 298K) for 10 mL catalyst:

GHSV = (mL feed/hr)/mL cat

mL feed/hr = GHSV * (mL cat) = (280 mL feed/mLcat-hr) (10 mL cat) = 2800 mL feed/hr

2800 mL feed/hr * (1 hr/60 min) = 46.7 mL/min at 698K

46.7 (298/698) = 20 mL/min at 298K (25C)

h. At a flow rate of 20 mL/min (298K), mL catalyst for 8.6 sec contact time:

20 mL/min at 298K = 46.7 mL/min at 698K

8.6 sec = 0.143 min

contact time = mL cat/(mL feed/min)

mL cat = (contact time)*(mL feed/min) = (0.143 min)*(46.7mL/min) = 6.8 mL

i. At 30 mL/min (298K), mL of catalyst for 4.3 sec contact time:

30 mL/min at 298K = 70.3 mL/min at 698K

4.3 sec = 0.072 min

contact time = mL cat/(mL feed/min)

mL cat = (contact time)*(mL feed/min) = (0.072 min)*(70.3/min) = 5.1 mL

$GHSV = (1/4.3) \text{ sec}^{-1} = 0.232 \text{ sec}^{-1} = 13.95 \text{ min}^{-1} = 837 \text{ hr}^{-1}$

4. a-e

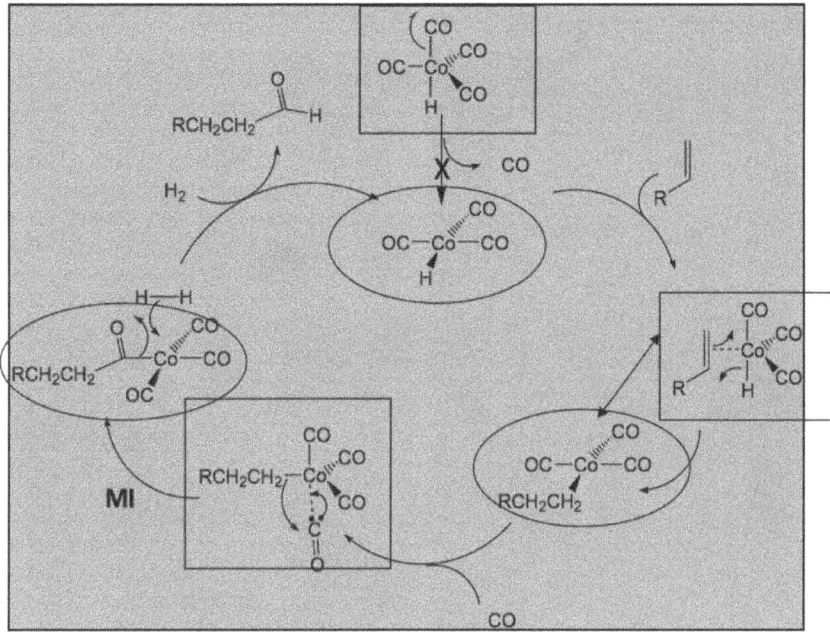

(f) RCH=CH$_2$ + CO + H$_2$ → RCH$_2$CH$_2$CHO

(g) HCo(CO)$_4$ is really the catalyst precursor, because, as represented in this scheme, it is outside the catalytic cycle.

(h) Homogeneous: single metallic center, ligand dissociation, coordination activation of olefin RCH=CH$_2$, ligand (CO) environment.

2

Acid Catalysis

Acid catalysts are among the most commonly used in industry for large scale production of both fuels and chemicals. They represent much of the catalyst technology that has been developed over the last 50 years. This chapter will cover aspects of theory and acid catalyst design (which are also common to a number of catalyst types), including surface area, catalyst supports, and thermal treatment and phase formation. A fundamental understanding of these aspects will form the basis for the study of the other catalyst types that will be covered in subsequent chapters. Elements of this chapter follow a pattern common to all subsequent chapters. These include a discussion of the major elements used (a "catalytic chemist's periodic table"), thermodynamic issues, basic reaction steps, catalyst synthesis and commercial examples.

An important aspect in understanding the chemistry of these systems and their limiting factors is the mechanism by which these reactions occur. Thus, a critical feature of this as well as subsequent chapters is a presentation of the major catalytic mechanisms. In most cases, the major elements of the mechanism have been elucidated, but for some, only certain features of the mechanism are known. In those latter cases, a mechanism consistent with those known features is "proposed," and is so identified.

The industrial processes presented in Secs. 2.3 and 2.4 provide representative examples of high-volume commercial applications of acid catalysts and the mechanisms by which they operate. These are divided into refinery (Sec. 2.3) and chemical (Sec. 2.4) uses. A brief perspective into future trends and how acid catalysts could help provide practical solutions to problems, and opportunities for new products in the chemicals and fuels industries, is also discussed.

2.1 Basic Concepts

Acids are, by definition, proton sources (Bronsted acids) or electron poor materials (Lewis acids, which are electron acceptors). These attributes are reflected in the elements which are frequently used in acid catalysts.

2.1.1 *An Acid Catalyst Periodic Table*

A view of the periodic table elements employed in acid catalysts is typically:

- Hydrogen (as H^+),
- Transition metals in high oxidation states,
- Main group elements in the upper right corner or any halogen (that is, high electronegativity elements or those whose conjugate acids are strong acids).

Table 2.1.1:   Acid Catalyst Periodic Table

1a	2a	3b	4b	5b	6b	7b		8		1b	2b	3a	4a	5a	6a	7a	0
H																	He
Li	Be											B	C	N	O	F	Ne
Na	Mg											Al	Si	P	S	Cl	Ar
K	Ca	Sc	Ti	V	Cr	Mn	Fe	Co	Ni	Cu	Zn	Ga	Ge	As	Se	Br	Kr
Rb	Sr	Y	Zr	Nb	Mo	Te	Ru	Rh	Pd	Ag	Cd	In	Sn	Sb	Te	I	Xe
Cs	Ba	La*	Hf	Ta	W	Re	Os	Ir	Pt	Au	Hg	Tl	Pb	Bi	Po	At	Rn
Fr	Ra	Ac**															
*Lanthanides	Ce	Pr	Nd	Pm	Sm	Eu	Gd	Tb	Dy	Ho	Er	Tm	Y	Lu			
**Actinides	Th	Pa	U	Np	Pu	Am	Cm	Bk	Cf	Es	Fm	Md	No	Lw			

Common (yellow) catalytic elements.

Table 2.1.2:   Acid Catalysts

1a	2a	3b	4b	5b	6b	7b		8		1b	2b	3a	4a	5a	6a	7a	0
H												B	C	**O**			He
																	Ne
K	Ca	Sc	Ti	V	Cr	Mn	Fe	Co	Ni	Cu		**Al**	**Si**				
Rb	Sr	Y	Zr	Nb	Mo	Te	Ru	Rh	Pd	Ag							
Cs	Ba	La*	Hf	Ta	W	Re	Os	Ir	Pt	Au							
Fr	Ra	Ac**															
*Lanthanides	Ce	Pr	Nd	Pm	Sm	Eu	Gd	Tb	Dy	Ho	Er	Tm	Y	Lu			
**Actinides	Th	Pa	U	Np	Pu	Am	Cm	Bk	Cf	Es	Fm	Md	No	Lw			

The specific elements that dominate the compositions of industrial catalysts are H, as H^+ (Bronsted acids) and Si/Al/O, as silica, alumina and a variety of silica–alumina containing materials such as zeolites and clays. The high activity and low cost of these systems have resulted in their extensive study and the development of very efficient acid catalysts which employ these elements.

2.1.2 *pKa Acidity Scale*

The conventional measure of the strength of an acid BH is its pKa, which is the $-\log$ (K_{eq}) for Eq. 2.1.2.1 as defined by Eq. 2.1.2.2. This is most commonly measured in aqueous media by simple pH measurement. In this case, the actual chemical equilibria is Eq. 2.1.2.3, where a proton from the acid BH is donated to water to form an hydronium ion (H_3O^+), which has a pKa of about -1. For acids with pKa,s in this same order of magnitude, this works well. Examples of pKa,s of acids, which are commonly used as acid catalyists are shown in Table 2.1.2.

For very strong acids with pKa's < -12 (the acid strength of 100% sulfuric acid), the equilibrium of Eq. 2.1.2.3 is essentially completely to the right, and it becomes impossible to distinguish acid strengths or measure pKa in water, because [HB] is effectively zero. For these "superacids" a new measure of acidity is needed which uses a media more weakly basic than water.

Table 2.1.2:   pKa Acidity Scale (Ref. 8)

(2.1.2.1)

pK_a for the acid BH
$= -\log K_{eq}$ for the equilibrium:
$$HB \rightleftharpoons B^- + H^+$$
$$K_{eq} = [H^+][B^-]/[HB]$$

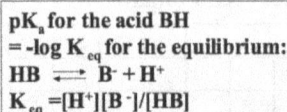

pKa's of Common Acid Catalysts:

Acid	pKa
H2SO4 (100%)	-12
H2SO4 (conc)	-1
HCl (conc), HNO3	-1
CF3CO2H	1
-Si-OH (surface)	2
RCO2H	5
RNH3+	5
PhOH	10
H2O	16

(2.1.2.2)

- pK_a is measured in H_2O, so
  $$HB + H_2O \rightleftharpoons B^- + H_3O^+$$
- $pK_a (H_3O^+) = -1$
- If pK_a (BH) $\ll$ -1, [HB] $\sim$ 0, and so K_{eq} cannot be measured.
- Therefore, a new acidity scale needed for acids with $pK_a \ll$ -1

(2.1.2.3)

2.1.3 *Ho Acidity Scale for Superacids*

The Hammett acidity scale was developed to measure the acid strength of superacids (pKa<−12). This uses a non-aqueous media and a series of aniline base indicators of known acid strength, known as Hammett bases. The pKa's for this series of Hammett aniline base indicators (B) has been measured using Eqs. 2.1.3.1–2.1.3.3. These bases were chosen to undergo a color change as [BH+]/[B] reaches a critical value, indicating the presence of BH+ and the donation of a proton from the superacid to indicator base B according to Eq. 2.1.3.4.

The Ho determination is made by adding a series of indicators of increasing basic strength (decreasing acid strengths of BH+) to a non-aqueous solution or dispersion of the superacid. At the point where a color change occurs, Ho can be calculated using the pKa of BH+ (the acid form of the indicator B) and the critical value of [BH+]/[B] at the point of color change. [BH+]/[B] can also be measured by spectroscopy. As for the pKa scale, the more negative the Ho value, the stronger the acid.

Hammett acidity method can be applied to both liquid and solid superacids. Ho values for a number of representative superacid catalysts (liquids in red and solids in blue) are shown in Fig. 2.1.3. These range from −13 for phosphotungstic acid to −20 for FSO_3H/SbF_5, which is one of the strongest superacids known.

Other examples of solid acid catalysts have been recently reviewed (Tanabe). Phosphotungstic acid has been extensively studied as a solid

(2.1.3.1)

Hammet Acidity (Ho)
• For Superacids (pK$_a$<-12)
• Non-aqueous media
• Uses Hammett Base Indicators
(series of aniline bases)

Keq's for a series of Hammett Bases
(B's) were measured (non-aqueous):
BH+ ⇌ H+ + B
(2.1.3.2)
K(BH+) = [B][H+]/[BH+]
pK(BH+) = -log ([B][H+]/[BH+])

When an acid is added to B, the Ho
value for that acid is defined as:
Ho= pK(BH+)-log([BH+]/[B]).
[BH+]/[B] can be observed by
spectroscopy or color change.

Ho for Some Superacid Catalysts

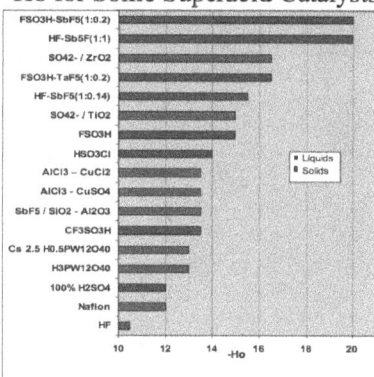

(2.1.3.3)

Fig. 2.1.3   Superacids (Ref. 8)

superacid and is discussed in more detail in Section 2.1.5.

2.1.4 *Superacid Example: Phosphotungstic Acid*

Phosphotungstic acid ($H_3PW_{12}O_{40}$) (Fig. 2.1.4) belongs to a group of inorganic mixed metal oxides known as "heteropoly-acids" (also known as "polyoxometallates" and "metal oxide clusters"). These materials have a solid state arrangement of atoms known as a "Keggin" structure with a central tetrahedral phosphate (PO_4) core surrounded by 12 WO_6 octahedra. The 12 octahedra can be divided into four groups, each with three edge-shared octahedral of the composition W_3O_{13}. Other common Keggin structures, which are also strongly acidic solids, contain Mo in place of W and Si in place of P.

Superacidity results from:

• Weakly Basic Anion
• Delocalized Negative Charge
• Oxidative and Thermal Stability

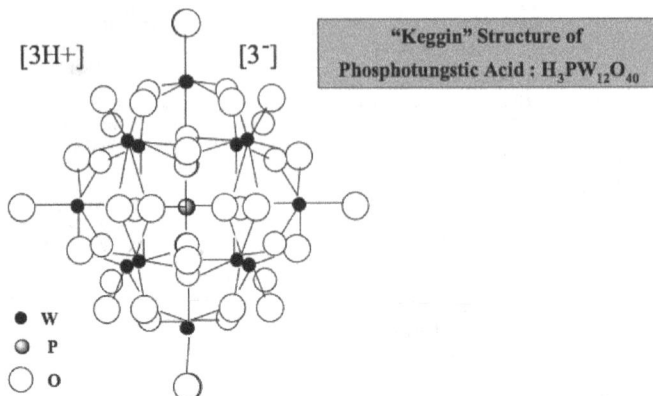

Fig. 2.1.4   Superacid Example: Phosphotungstic Acid (Ref. 9) reproduced with permission

   This arrangement of P, W and O atoms forms an anion with a −3 charge, which is spread out over many of the 40 oxygen atoms, and with three protons, which are associated with three of the bridging oxygen atoms. The delocalization of negative charge stabilized the anion, analogous to the resonance stabilization of negative charge in carboxylate (RCO_2), sulfate $(SO_4)^{2-}$ or phosphate $(PO_4)^{3-}$ anions. This accounts for the weak basicity of these anions and the strong acidity of the corresponding conjugate acids — carboxylic, sulfuric and phosphoric acids (RCO_2, H, H_2SO_4, H_3PO_4).

   In the case of the phosphotungstic acid, the negative charge is spread over many more oxygen atoms (than in the parent phosphoric acid, for example), resulting in a very "soft," or diffuse charge which is distributed over the very large surface of the Keggin anion. The W and P atoms in phosphotungstic acid are in their highest, most thermodynamically stable oxidation state, resulting in high oxidative and thermal stability. That is, it is not chemically altered by oxidizing agents (for example, air or oxygen) or by heat (up to about 400°C).

2.1.5 *Bronsted Versus Lewis Acidity*

Both Lewis and Bronsted acids are used as catalysts (or catalyst precursors). Bronsted (or protonic) acids activate a reactant by donation of a

$$\text{Lewis Acid} \xrightleftharpoons[\text{-H2O}]{\text{+H2O}} \text{Bronsted Acid} \qquad (2.1.5.1)$$

Homogenous Examples:

$$BF3 + H2O \rightarrow [B(OH)F3]- H+ \qquad (2.1.5.2)$$

Heterogeneous Examples:

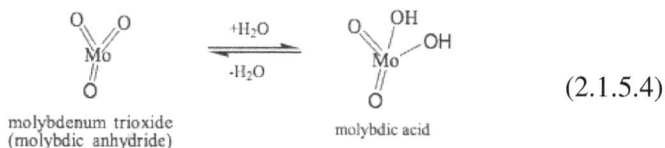

$$AlCl3 + H2O \rightarrow [Al(OH)Cl3]- H+ \qquad (2.1.5.3)$$

molybdenum trioxide   molybdic acid
(molybdic anhydride)

$$(2.1.5.4)$$

proton, while Lewis acids are an electron-poor species which do not actually possess a proton, but can perform catalytic activation by acceptance of an electron from a reactant.

Lewis acids can be converted, often reversibly, to the corresponding Bronsted acid by hydration or reaction with another protonic species, like an alcohol, according to Eq. 2.1.5.1. Therefore, the actual acidic species present, and thus the operative mechanism, depends on the presence (proton activation) or absence (Lewis acid activation) of a protonic species.

The conversion of boron trifluoride, a Lewis acid commonly used in industry as a homogeneous acid catalyst, to a Bronsted acid by reaction with water is shown in Eq. 2.1.5.1. The analogous reactions for aluminum trichloride ($AlCl_3$) and molybdenum trioxide (MoO_3), two heterogeneous Lewis acid catalysts, are shown in Eqs. 2.1.5.3 and 2.1.5.4.

2.1.6 *Some Common Solid Acid Catalysts*

As was discussed previously (see 1.7.4), solid heterogeneous catalysts are prevalent in industry because of their many practical advantages. Acid catalysts are no exception to this and some of the more commonly used solid acids which are used throughout industry are shown in Table 2.1.6.

Table 2.1.6:   Some Common Solid Acid Catalysts

- Silica/Alumina–Type Materials
  — SiO_2 (silica)
  — Al_2O_3 (alumina)
  — SiO_2/Al_2O_3 (silica–alumina)
  — Clays (e.g., Mordenite, Bentonite)
  — Zeolites
- Acidic Ion Exchange Resins (Sulfonated Polymers):
  — $(CF(SO_3H)CF_2)n$– Nafion®
  — $[(C_6H_4$–$SO_3H)CH$–$CH_2]n^-$
  — Amberlyst ®
- Acidified Solids: e.g., above treated with HF, H_2SO_4, H_3PO_4 or other strong liquid acids

These fall into three major categories:

- Silica–alumina
- Acidic forms of ion exchange resins (e.g., sulfonated polystyrene) and
- Acid-treated solids

The silica/alumina type materials are the most common inorganic solid acids, which can be further divided into the pure elemental oxides (silica, SiO_2 and alumina, Al_2O_3), mixed metal oxides (for example, synthetic silica–aluminas, or naturally-occuring clays), and crystalline silica–aluminas known as zeolites.

Among the typical organic-based solid acids are the acidic ion exchange resins, including sulfonated polymers such as the per-fluorinated sulfonated polyethylene resin with the trade name Nafion™ (made by DuPont) and the sulfonated polystyrene Amberlyst™ (made by Rohm and Haas) catalysts.

Inorganic acids and other commercially-available solids can be made acidic (or more acidic) by treatment with strong liquid acids. A specific example is a catalyst know as "solid phosphoric acid", or "SPA" catalyst, which is phosphoric acid adsorbed on a clay. In these materials, the catalytic species may be the liquid acid itself, which is strongly adsorbed to the surface, or a Bronsted acid site on the catalyst which is formed by reaction with the liquid acid (similar to Eq. 2.1.5.1).

2.1.7 *Silica–Alumina SiO_2/Al_2O_3 Solid Acids*

Silica and alumina-type materials are probably the most prevalent solid acid catalysts in commercial use today. Not only are they inexpensive, readily available, and very stable materials, but their acidic properties can be controlled by the ratio of silica to alumina in the solid state.

Pure silica is composed of tetrahedral SiO_4 units and AlO_3 alumina units, which, when pure, are trigonal. However, when normally trigonal alumina is added (or in solid state electronic terms, "doped") into a tetrahedral silica matrix, it adopts a tetrahedral AlO_4 — form and in so doing, creates a negatively charged Al atom. In order to compensate this negative charge in the solid, which is charge-neutral, these negatively charged centers become the sites for acidic protons, as shown in Fig. 2.1.7. Thus, these protonated oxygen atoms bridging Si and Al atoms are the acidic sites in silica–alumina catalysts.

The incorporation of tetrahedral alumina into the silica framework and the formation of active acidic sites requires more than just a physical mixing of silica and alumina. The monomeric tetrahedral SiO_4 and AlO_4 units must come into intimate contact to allow the formation of the bridging silica and alumina oxygen and the formation of the alumina-substituted silica matrix. One means of accomplishing this is by the addition of a soluble Al^{3+} salt to a suspension of such monomeric silica units known as a "silica hydrogel."

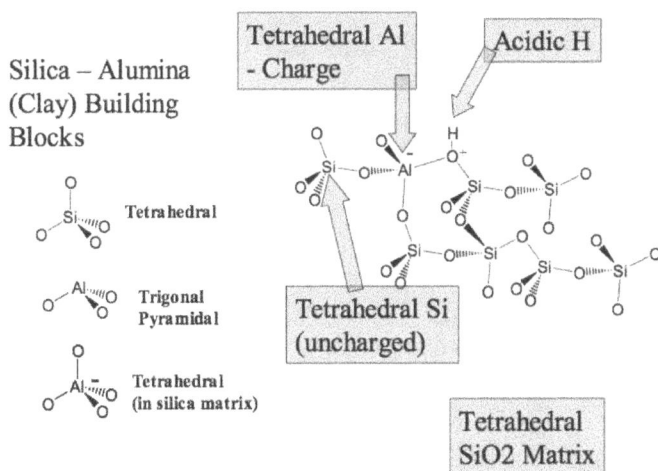

Fig. 2.1.7 SiO_2/Al_2O_3 Solid Acids (Ref. 5)

2.1.8 *Zeolite Synthesis*

Most silica–aluminas are amorphous materials, meaning that there is no long-range ordering of atoms to form a discrete crystallographic phase. This means that the range of Si:Al ratios possible in these catalysts is very broad and not limited by the phase diagram (a mapping of the boundaries of thermodynamically stable phases as a function of atomic composition).

However, there is a class of crystalline silica–alumina type materials, which have discrete crystalline structures and which have found extensive use as solid acid catalysts. These materials are known as zeolites and are formed in what is known as a hydrothermal process (involves water and heat) from sodium hydroxide, sodium aluminate and sodium silicate (also known as "water glass"). An aluminosilicate gel forms, which is subsequently converted to the crystalline zeolite (Eq. 2.1.8).

The structure of the zeolite which forms depends on the composition (mainly the Si:Al ratio which usually varies from about 1–10, but can be 100 or even higher), the temperature (usually around 100°C), the reaction time (which can be from a few hours to several days), and the pH (which is controlled by the sodium hydroxide). In order to obtain the pure desired zeolite, it must be washed to remove unreacted sodium silicate from the pores.

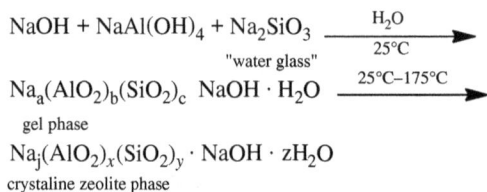

$$NaOH + NaAl(OH)_4 + Na_2SiO_3 \xrightarrow[25°C]{H_2O}$$

"water glass"

$$Na_a(AlO_2)_b(SiO_2)_c \; NaOH \cdot H_2O \xrightarrow{25°C–175°C}$$

gel phase

$$Na_j(AlO_2)_x(SiO_2)_y \cdot NaOH \cdot zH_2O$$

crystaline zeolite phase

2.1.8. Zeolite Synthesis (Ref. 5)

2.1.9 *Zeolite Acid Sites*

Because sodium salts and caustic are used in zeolite synthesis, the Na^+ salts of the zeolite, not the acidic form, are obtained as it is prepared (in Eq. 2.1.8). In order to convert these to the acid form, the sodium form is first treated with an ammonium salt to convert it to the ammonium-exchanged form, followed by heating at about 350°C to liberate ammonia

Zeolite as prepared:

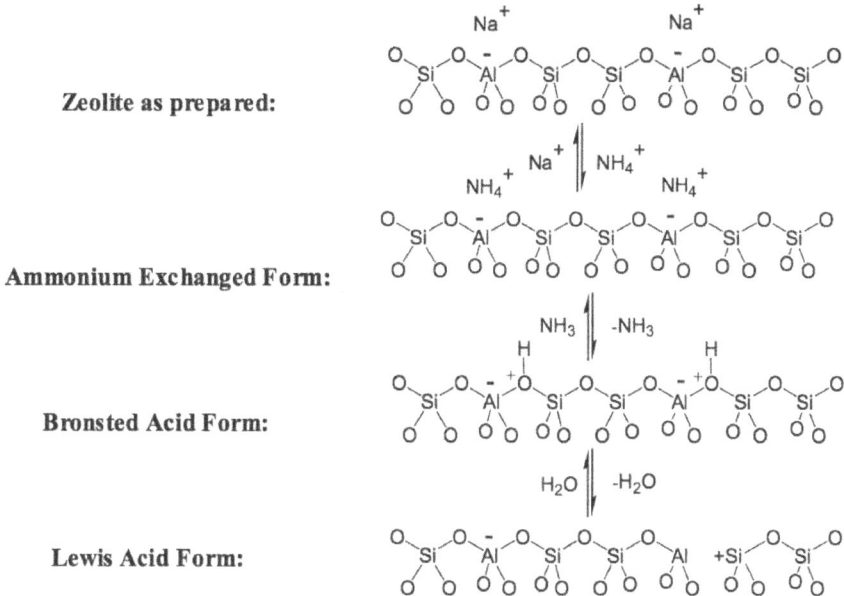

Ammonium Exchanged Form:

Bronsted Acid Form:

Lewis Acid Form:

Fig. 2.1.9 Zeolite Acid Sites (Refs. 5, 10)

and form the protonic (or Bronsted acid) form of the zeolite. Further heating at about 450°C liberates water from the structure to form the Lewis acid form of the zeolite.

The loss of water from the Bronsted form is reversible and thus, it can be reformed by adding water at between 350°C and 450°C. From the stoichiometry of the equations shown in Fig. 2.1.9, it can be seen that two Bronsted sites are required to generate one Lewis acid site.

2.1.10 *Bronsted and Lewis Acidity Measurements in Solids*

Bases such as pyridine can be used to probe Bronsted and Lewis acid sites in solid acids. The conversion of Bronsted to Lewis sites by *heat treatment* (also known as *calcination*) can be followed by adsorption of pyridine onto the zeolites and measuring the pyridinium ion (PyrH$^+$) adsorption (1550 cm^{-1}) on Bronsted acid sites versus the Lewis–acid coordinated pyridine (1450 cm^{-1}) (Fig. 2.1.10.1). The zeolite calcined at 350°C shows mainly Bronsted adsorption while the one at 800°C shows mostly Lewis sites.

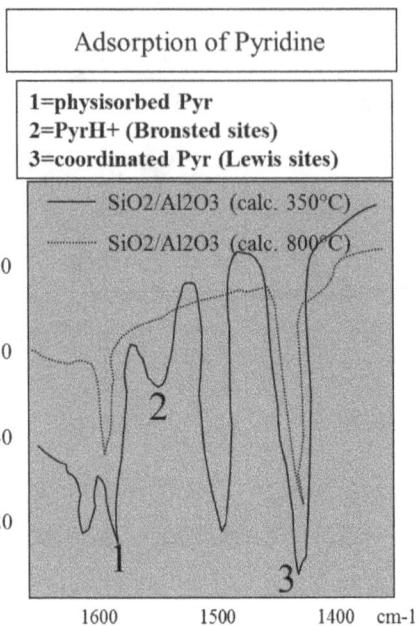

Fig. 2.1.10.1   **Bronsted and Lewis Acidity Measurement in Solids** (Ref. 11)

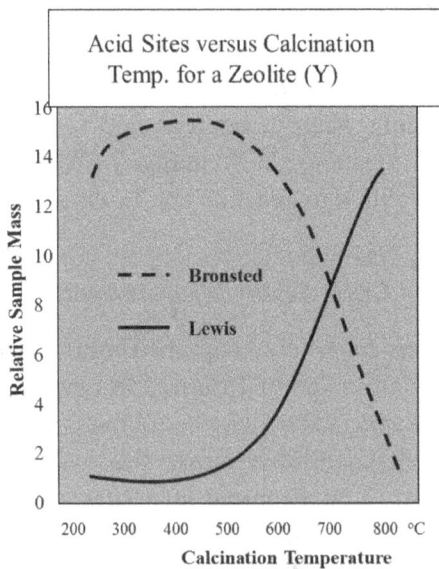

Fig. 2.1.10.2   **Acid Sites in Zeolites as a Function of Calcination Temperature** (Refs. 5, 10)

2.1.11 *Mechanism of Acid Catalyzed Aromatic Reactions*

Silica–alumina solid acids catalyze the isomerization, transalkylation and cracking (dealkylation) reactions of alkyl aromatics. The basic reaction mechanism by which silica–alumina-based solid acids operate is illustrated in Scheme 2.1.11 for the activation of olefins or aromatics. The Bronsted acid site of a silica–alumina catalyst. $[-O-Si-(OH)^{+}-Al-O]$ serves as a proton donor to the $C = C$ double bond (Fig. 2.1.11(a)) or to an aromatic ring (Fig. 2.1.11(b)) to form the corresponding carbenium ions.

These two reactions are the basic steps in the conversions of Scheme 2.1.11. An alkyl aromatic can be reversibly protonated to form the cyclohexadienyl carbenium ion (II) which can then reversibly lose an alkyl carbenium ion (I) to form the neutral aromatic, which upon loss of a proton forms the olefin. So the acid catalyst provides a pathway for the formation of all the intermediates required for cracking (II), isomerization (I) and transalkylation (I). This will be discussed in more detail in the context of specific industrial applications (for example, in Secs. 2.2.1 and 2.3.2).

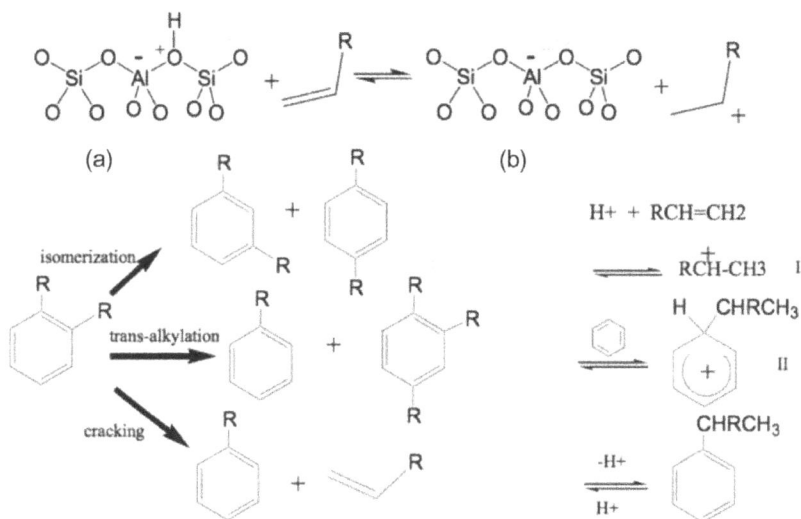

Fig. 2.1.11 **Mechanism of Acid Catalyzed Aromatic Reactions** (Ref. 12)

2.1.12 *Thermodynamics*

The thermodynamic limitations of catalytic reactions were discussed in 1.4, but it is worth repeating these in the context of acid catalysis. Recall that a catalyst cannot alter thermodynamics. It can only accelerate the rate of formation of a desired reaction product (the selective one) versus others. The formation of the most thermodynamically favored product will usually be a major concern, since the desired selective product is rarely the most thermodynamically favored one.

The thermodynamic state of a hydrocarbon depends on whether oxygen (air) or hydrogen is present (Fig. 2.1.12). In the presence of hydrogen, the most stable hydrocarbon is methane which is the product of hydrogen addition with C–C bond scission (a process called hydrogenolysis). In the presence of air or oxygen, the most stable product is CO_2 and water which results from total combustion. In the absence of hydrogen or oxygen (or any other oxidizing or reducing agents), the most stable state of a hydrocarbon is carbon (also known as coke), which results from loss of hydrogen.

Since most acid-catalyzed reactions do not involve oxidation or reduction, the major non-selective product is coke formation. Coke is very problematic in acid-catalyzed reactions, not only because it is an undesired product that consumes reactant (hydrocarbon), but also because it can poison many acid catalysts by covering surface sites and plugging catalyst pores. This will be discussed further in subsequent sections of this Chapter (for example, Sec. 2.2.3).

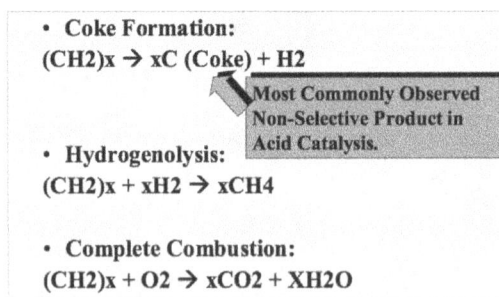

- **Coke Formation:**
  (CH2)x → xC (Coke) + H2

  Most Commonly Observed Non-Selective Product in Acid Catalysis.

- **Hydrogenolysis:**
  (CH2)x + xH2 → xCH4

- **Complete Combustion:**
  (CH2)x + O2 → xCO2 + XH2O

Fig. 2.1.12   Thermodynamic Products of Hydrocarbon Reactions

2.2 Elements of Acid Catalyst Design

Section 2.2 deals with the application of the concepts of acid catalysis presented in 2.1 to the design of acid catalyst for a desired reaction.

2.2.1 *Active Site Distribution*

The distribution of Bronsted and Lewis sites in a solid acid catalyst can have a dramatic effect on the resulting product distribution. This distribution can be controlled by the dehydration of Bronsted sites by heat treatment to form Lewis sites.

For example, the rate of isobutane cracking to propylene and methane (Fig. 2.2.1.1) is proportional to the number of Lewis acid groups, consistent with the active sites for this reaction being Lewis acid sites. This suggests that the reaction is initiated by hydride abstraction Lewis Acid sites (See Sec. 2.3.1).

The isomerization of an ortho di-alkylbenzene (Fig. 2.2.1.2), however, increases with the number of Bronsted acid sites, suggesting that proton

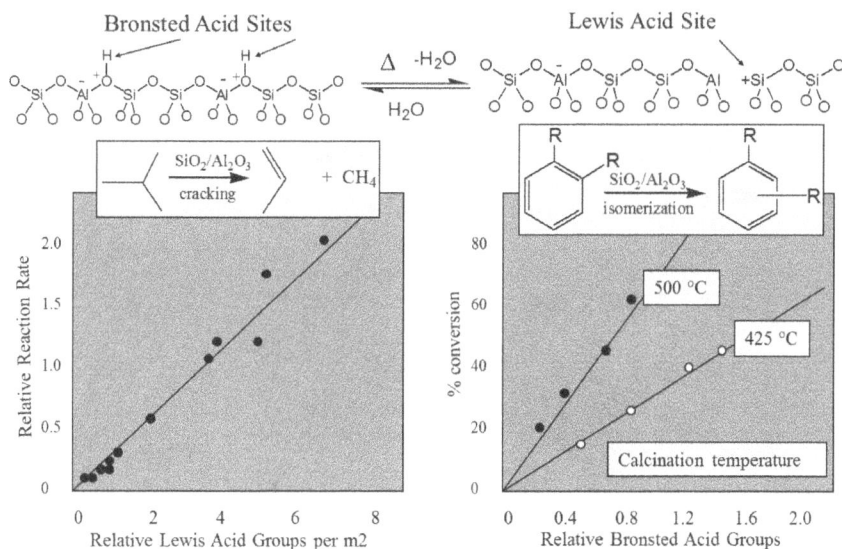

Fig. 2.2.1.1 Active Site Distribution in Cracking (Ref. 5)

Fig. 2.2.1.2 Active Site Distribution in Isomerization (Ref. 5)

transfer from a Bronsted site to the aromatic is the mechanism for this reaction.

2.2.2 *Zeolites*

As was mentioned previously (Sec. 2.1.8), zeolites are a special class of crystalline silica–aluminas. Zeolites are formed by assembly of tetrahedral SiO_4 or $(AlO_4)^-$ units, first into a building block composed of 24 tetrahedral units known as a sodalite unit, and then of sodalite units into a zeolite "superstructure." The structure shown in Fig. 2.2.2 represents the superstructure known as a *faujacite*, where Si of Al atoms are at the corners of the intersecting lines and O atoms are inbetween the Si or Al atoms.

Within the superstructure of the zeolite are cavities with openings, the size of which depend on the way in which the sodalite units are assembled. The size of the cavities and the openings determine the size of the molecules that can be adsorbed into (or desorbed from) and that can react within the zeolite "cage". In the faujacite superstructure, the large cavity is 12 Å in diameter and is surrounded by 6 sodalite units, forming an opening that is 7.4 Å in diameter. These dimensions are large enough to

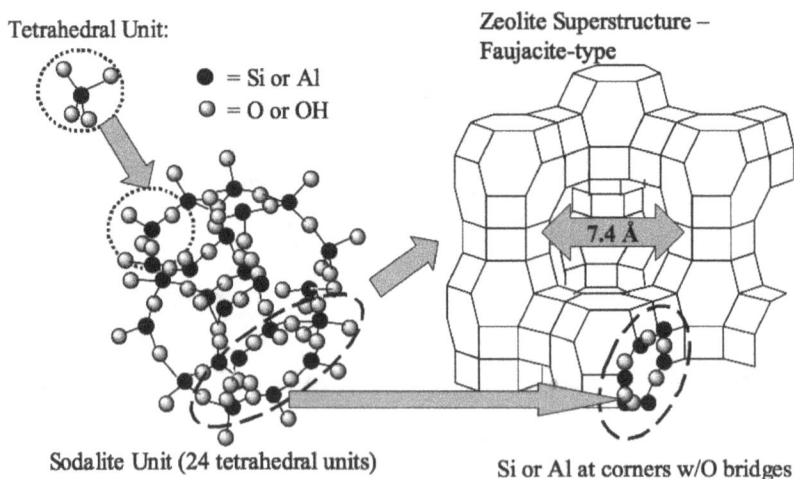

Fig. 2.2.2 Zeolites (Ref. 13, reproduced with permission)

accommodate many organic molecules such as small alkyl aromatics and short hydrocarbons.

2.2.3 *Other Si/Al Frameworks*

Other assemblies of sodalite units include sodalite, in which the sodalite units are directly stacked on one another, and zeolite A, in which the stacks are separated by an oxygen layer in one dimension (Fig. 2.2.3.1). In sodalite, the diameter of the cavities is only 2.6 Å, and so only very small molecules like water and methanol can be adsorbed. Zeoilite A has 4.2 Å diameter cavities, which can adsorb somewhat larger molecules than sodalite, but smaller ones than faujacite.

Other zeolite frameworks include mordenite and ZSM-5, the cross sections of which are shown in Fig. 2.2.3.2. Mordenite has straight 8 Å (max free dia, Refs. 10, 12, 14) channels, which can accommodate many organic reactants. It has highly active surface sites, but is prone to channel plugging by coke. Since there are no other intersecting channels in the structure, once a pore is plugged by coke or polymeric material, the accessibility of reactants to and the diffusion of products from the pore is cut off, thus deactivating the catalyst.

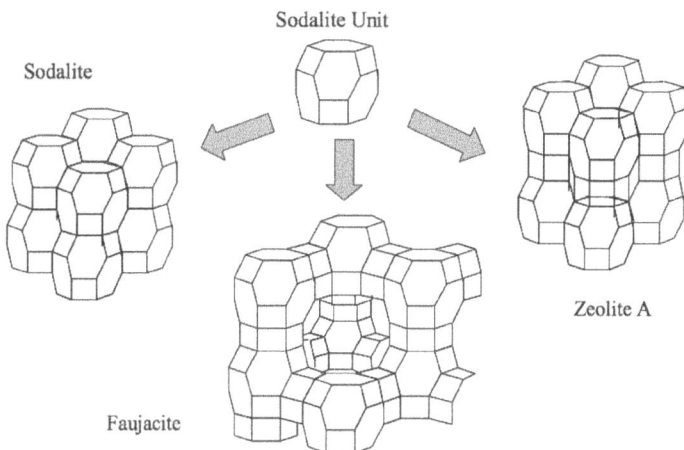

Fig. 2.2.3.1   Other Si/Al Frameworks (Ref. 5)

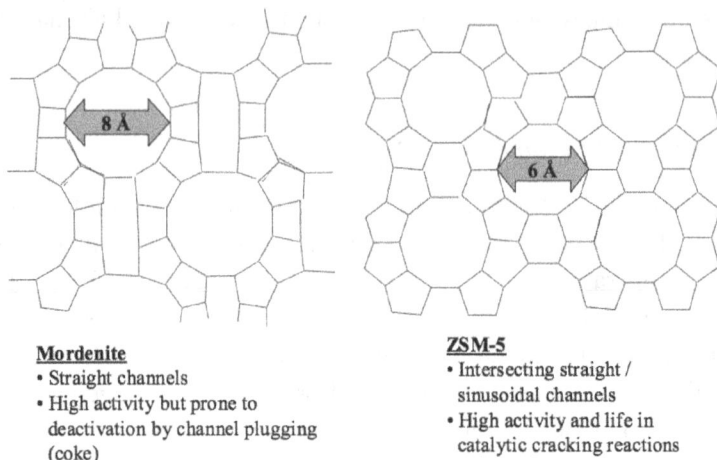

Mordenite
- Straight channels
- High activity but prone to deactivation by channel plugging (coke)

ZSM-5
- Intersecting straight / sinusoidal channels
- High activity and life in catalytic cracking reactions

Fig. 2.2.3.2   Other Si/Al Frameworks (Refs. 5, 13)

The ZSM-5, on the other hand, has intersecting straight and sinusoidal channels, which have high activity, and is much less susceptible to deactivation by coke. Polymeric structures, which form on other acid catalysts from smaller organic molecules and are the precursors to coke, do not form on zeolites because their bulky transition states are too large to be accommodated within the cavity (6 Å max free fid, Refs. 10, 12, 14). This is an example of shape selective by restricted transition state. Because of their resistance to coking, ZSM-5 has found extensive use as a cracking catalyst in refineries (Sec. 2.3.1).

2.2.4 Shape Selectivity

The ability of a catalyst to preferentially form products based on the less space-filling reactants, transitions states or products is known as "shape selectivity". Shape selectivity can be manifested by three mechanisms:

Reactant selectivity: the less bulky reactant diffuses into the cavity faster and therefore, reacts faster. For example, the hydrogenation of 1-hexene occurs 100 times faster than 4,4-dimethylhexene over Pt/ZSM-5.

Restricted transition state selectivity: The reaction involving the more bulky transition state cannot be easily accommodated within the zeolite cavity and is therefore slower (for example, inhibition of coke formation).

Product shape selectivity: Isomers equilibrate within the cavity and the less bulky one diffuses out of the opening faster, for example, toluene alkylation to form predominantly para alkylate.

The acid-catalyzed isomerization and transalkylation of 2-ethyl toluene (**2a**, Fig. 2.2.4.1) illustrates the shape selectivity of mordenite when compared to other acid catalysts. A measure of the isomerization conversion of **2a** in Eq. 2.2.4 is the sum of the two isomers which are formed (**2b** + **2c**) divided by the total isomers (**2a** + **2b** + **2c**). This ratio is plotted as the "reaction coordinate" on the *x*-axis in Fig. 2.2.4.1, which, at equilibrium, is close to 100%.

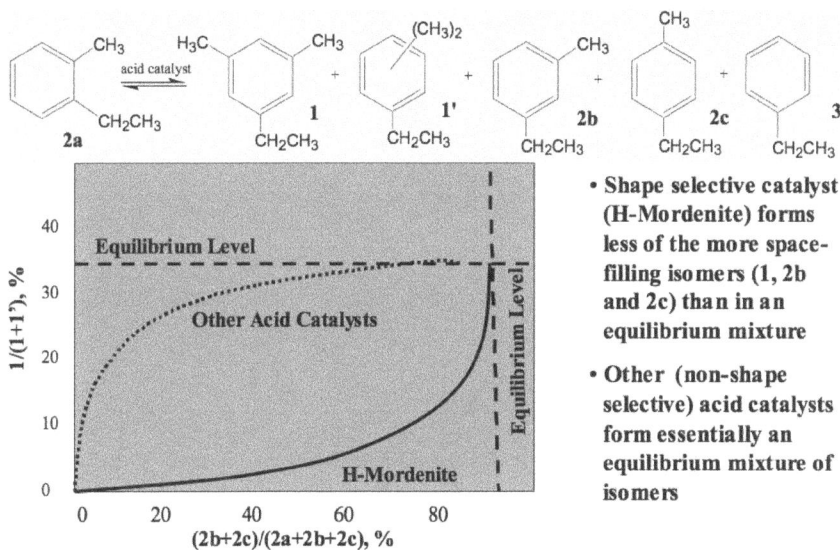

Fig. 2.2.4.1   Shape Selectivity (Ref. 14)

Likewise, the conversion of **2a** by trans-alkylation (transfer of a methyl group from one to another molecule of **2a**) can be followed by its conversion to the two trisubstituted isomers 1, 1′ and the monosubstituted isomer **3**. When this reaction occurs inside the cavity of a shape selective catalyst, **1** (the more space-filling isomer) will not fit through the cavity opening in the zeolite as easily as **1′** (the less space-filling isomers). Thus the ratio **1/(1 + 1′)** will be smaller for a selective than for a non-shape selective catalyst, which will approach the equilibrium ratio of about 35% as the overall conversion of **2a** increases, both types of catalysts will approach this same equilibrium level at 100% conversion of **2a**.

The space-filling capacity of the di and trisubstituted isomers is best shown using a three-dimensional (3D) molecular modeling program, such as is in ChemBioDraw, but, for simplicity here, is illustrated in Fig. 2.2.4, by simple circumscription of the molecular structures. The 5-ethyl 1,3-dimethyl isomer (**1**) is more space-filling than the 5-ethy l 1,4-dimethyl isomer (**1′**),

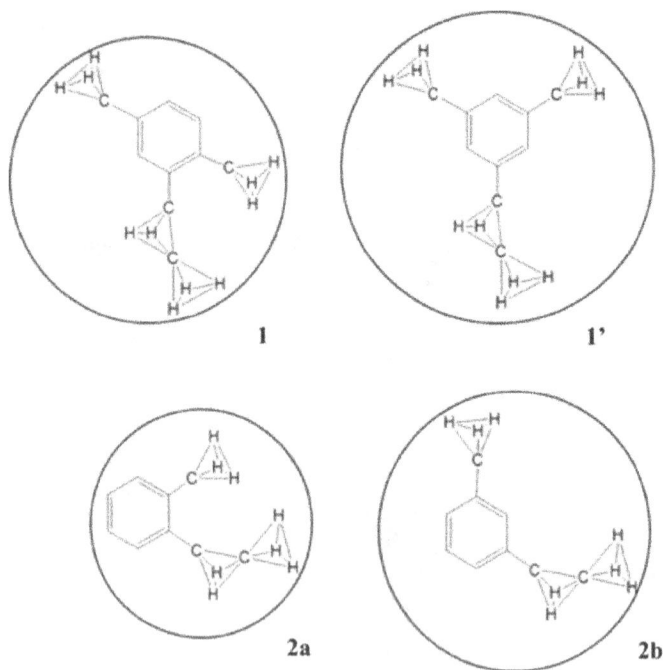

Fig. 2.2.4.2   Shape Selectivity (Refs. 10, 12, 14)

and both are more space-filling than the ethyl toluene isomers (**2a** and **2b**).

Thus, by running a shape selective catalyst at somewhat less than total conversion of **2a**, one can achieve a greater yield of **1′** than is otherwise possible from a non-space selective catalyst. This strategy is used in refining and in chemicals manufacture to enhance the production of a less space-filling isomer than is otherwise possible, or to remove a less space filling molecule by its selective adsorption into a shape-selective material.

2.2.5 *Surface Activity Effects*

As with other heterogeneous catalysts, the solid surface and particle properties of solid acids are an important aspect related to catalytic activity. Important measures of these properties include surface area, particle size and crystallite size. A catalyst support (i.e., an inert material on which the catalyst is adsorbed) can be used to tailor these properties. Surface area is the amount of catalyst surface available for reaction, usually expressed in m^2/g and measured by adsorption of N_2 (BET method, to be discussed in Sec. 7.2.1.)

The internal pores of a catalyst, not its geometric area, account for the vast majority of its surface area. The smaller the pores, the greater the surface area and total pore volume, but small pores are less accessible to reactants, and more suseptable to coking and diffusion limitations. Larger pores are more accessible to reactants, but this lowers total surface area and pore volume (Sec. 7.2.2). Small particles expose more catalyst surface

2.2.5 Surface Activity Relative Effects

Surface Area: Measures the catalyst surface available for molecular adsorption. (using N_2 adsorption — BET Method) Sg, m^2/g
Pore Volume: V_g, mm^3/g
Pore Size: P_d, mm

Particle Size: Expressed as an average or distribution of particle sizes, usually in μ.

Crystallite Size: The size of primary particles (aggregates of repeating crystallographic phase units) as measured by electron microscopy.

Catalyst Support: A method for increasing surface area, and dispersing an active phase on an inert material such as silica or alumina.

and pores than larger particles, but the former are harder to separate from products. Thus, there is a compromise between surface area, pore volume, pore size distribution and particle size to achieve optimal catalyst performance.

Macroscopic particles are made up of aggregates of crystallites (i.e., the smallest primary particle measurable by electron microscopy — to be discussed in Sec. 7.1.7), the size of which indicate the density of exposed catalytic sites. In general, the smaller the primary particles are, the more active is the catalyst.

One method of controlling these properties is to adsorb a solution of the active catalyst or catalyst precursor on a support which has the desired surface and particle properties. Many of the common acid catalysts such as silica- and alumina-containing materials are also supports for other active catalytic elements such as noble metals.

2.2.6 *Surface Activity: Small Crystallite Example*

As an example of the profound effect that surface area has on a catalyst is provided by the alkylation of trimethylbenzene by cyclohexene over a series of heteropolyacid catalysts of the formula $Cs_xH_{3-x}PW_{12}O_{40}$. While the Hammett acidity (Ho values) is about the same (Ho = −13) for all catalysts in the range $0 < x < 3$ (Fig. 2.2.6.1, the surface areas are greatest in the range $2.5 < x < 3.0$. The surface acidity of this series is proportional to the surface area and acid strength. As one moves from $x = 0$ to $x = 2$, the surface area, and thus, the surface acidity and the activity for the reaction decrease. Where $x = 3$ (the fully neutralized salt) there is no acidic proton, and thus, the surface acidity and the activity for the acid-catalyzed reaction shown is zero. The catalyst of composition $x = 2.5$ combines a very high surface area with residual acidic protons necessary to catalyze the reaction, and is therefore the catalyst composition with the highest activity.

The high surface area of the $x = 2.5$–3.0 salts results from the formation of small primary crystallites, which are about 8 nm in diameter (Fig. 2.2.6.2). This results in very small micropores (0.6–0.7 nm in diameter as measured by electron microscopy), and thus, a very high exposed surface area available for reaction which, in turn results in high surface acidity and high activity for the acid-catalyzed alkylation reaction in Fig. 2.2.6.1.

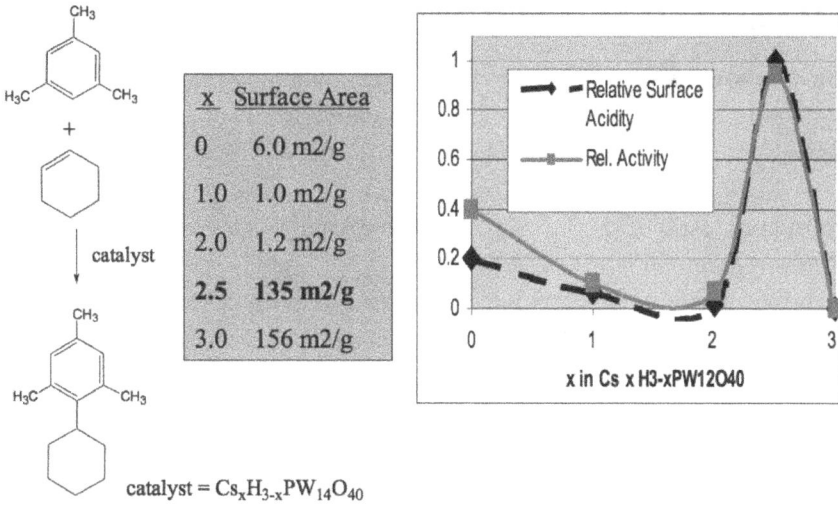

Fig. 2.2.6.1 **Surface Activity — Small Crystallite Example** (Refs. 8, 15)

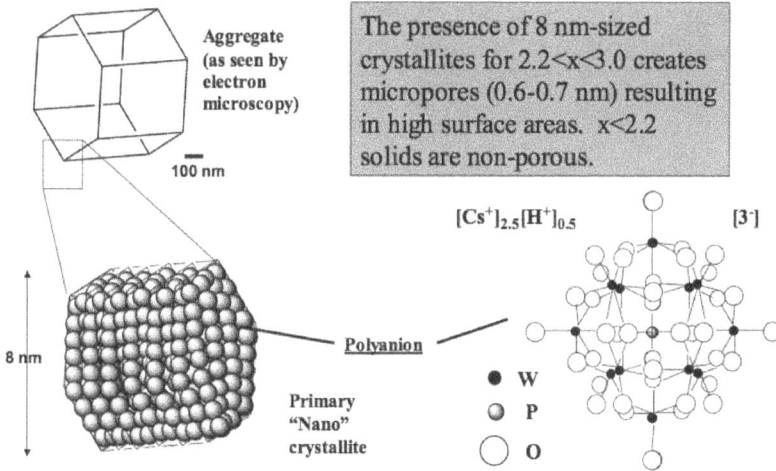

Fig. 2.2.6.2 **Surface Activity — Small Crystallite Example** (Refs. 8, 15 $Cs_{2.5}H_{0.5}$ $PW_{12}O_{40}$ structure adapted with permission from Ref. 16)

2.2.7 Heat Treatment — Phase Formation

We saw how heat treatment is important for obtaining the desired level of hydration, and thus for controlling the relative levels of Lewis–Bronsted sites (2.2.1). It is also an important step in the formation of the desired catalytically active phases. Consider again the case of the Cs salts of phosphotungstic acid (Fig. 2.2.7.1). These catalysts are made by adding increasing amounts of CsOH to $H_3PW_{12}O_{40}$. The Cs_0 and Cs_3 materials form the corresponding homogeneous phases with these compositions. For the $Cs_{1.0}$, $Cs_{2.0}$ and $Cs_{2.5}$ materials, the phases formed in the slurries are mixtures of the Cs_0 and $Cs_{3.0}$ phases. It is only after heat treatment that the most active $Cs_{2.5}$ phase forms (as a homogeneous phase for the $Cs_{2.5}$ composition, and mixed with the Cs_0 phase in the $Cs_{1.0}$ and $Cs_{2.0}$ compositions).

Thus, we have now seen that heat treatment is an essential step in catalyst synthesis for controlling:

— Hydration levels;
— Lewis versus Bronsted acidity and
— Active catalyst phase formation.

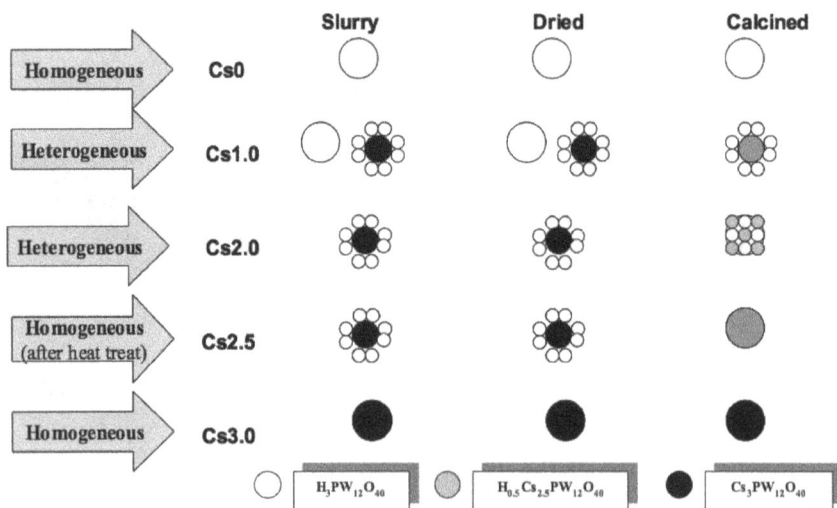

Fig. 2.2.7.1 Catalyst Heat Treatment — Phase Formation (Ref. 15)

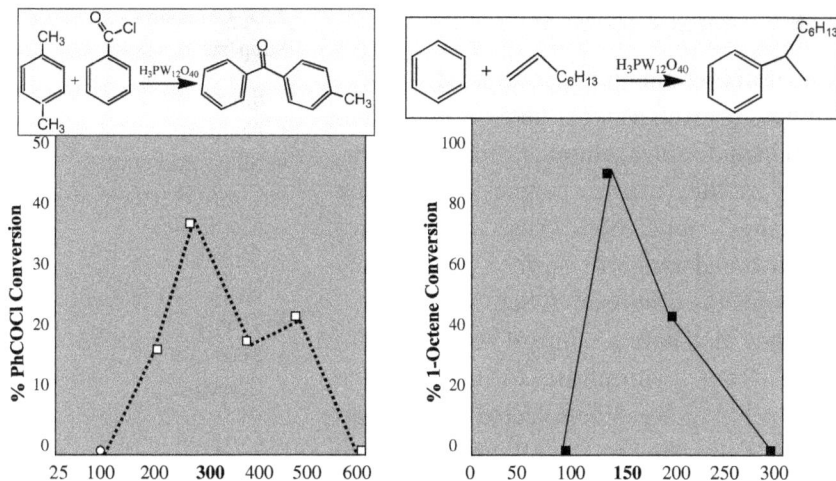

Fig. 2.2.7.2   Heat Treatment — Affect on Selectivity (Ref. 17)

Heat treatment also controls surface area. Higher heat treatments (generally above 400°C) reduced surface area by thermal fusing of particles (by a process known as *sintering*) and collapsing of pore structure. Heat treatment also removes extraneous impurities which volatilize and/or oxidize (heat treatment in air or oxygen) to gaseous products, thus eliminating them from the catalyst.

Another example of how heat treatment controls activity for Lewis versus Bronsted acid-catalyzed reactions is shown in Fig. 2.2.7.2. The acylation of toluene with benzoyl chloride (Fig. 2.2.7.2.1) occurs on Lewis acid sites (possibly via chloride abstraction) and is thus, favored by higher heat treatments (300–500°C). The alkylation reaction, on the other hand (Fig. 2.2.7.2.2), requires protonation of olefin and therefore is favored by Bronsted sites and lower heat treatment (150°C). (More details on the mechanism of aromatic alkylation are shown in Fig. 2.4.1).

2.2.8 *Common Catalyst Supports*

Among the common supports used for acid catalysts (as well as other heteroeneous catalysts) include silica, alumina, silica–alumina, clays, zeolite, titania, magnesia (also has basic sites) and carbon. These supports provide a broad range of surface and particle properties (as discussed in Sec. 2.2.5).

They also have a range of surface chemistries from hydrophilic to hydrophobic character, which can be selected for compatibility with the hydrophilic:hydrophobic character of the active catalyst to be supported. These effects are discussed in more detail in Sec. 2.2.9.

2.2.9 *Catalyst Supports: Hydrophobic Versus Hydrophillic Surfaces*

The two extremes of surface polarity are represented by silica (hydrophilic) and carbon black. The latter is very sparsely populated with polar (oxygenated, acidic) groups and is thus highly hydrophobic. Catalysts are easily adsorbed onto carbon from non-polar media. On the other hand, silica is bristling with surface –OH groups, resulting in a highly acidity and hydrophilic surface with about 500 times the level of strong acids sites as carbon black.

We have now seen how the selection of the proper catalyst support with the optimal surface area, pore and particle properties, and surface polarity can greatly improve the activity of an acid catalyst, as well as any heterogeneous catalyst.

2.2.8 Common Catalyst Supports

- Silica
- Alumina
- Silica–Alumina
- Clays (Bentonite, Kaolin, Mordenite)
- Zeolites (A, X, Y, Faujacites, ZSM-5)
- Titania
- Magnesia
- Carbon

Carbon Black vs. Silica

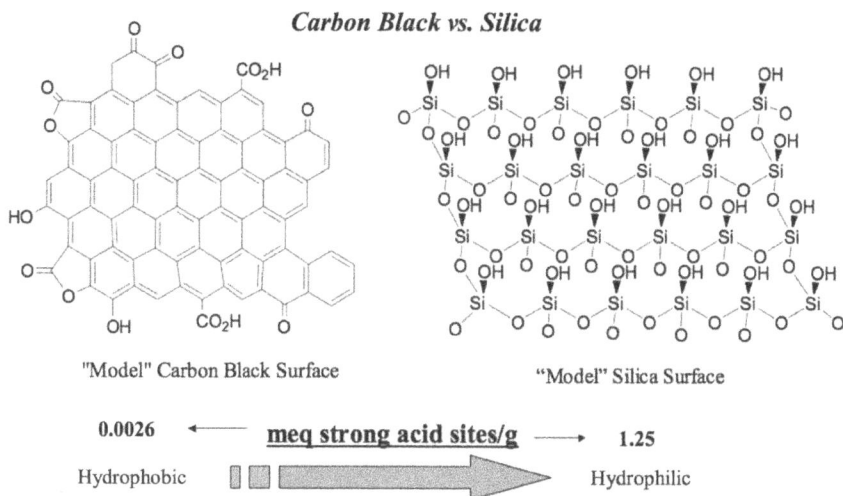

"Model" Carbon Black Surface   "Model" Silica Surface

0.0026 ◀——— **meq strong acid sites/g** ———▶ 1.25

Hydrophobic   Hydrophilic

Fig 2.2.9   Hydrophobic Versus Hydrophilic Surfaces

2.3 Major Industrial Processes — Refinery

Major commercial applications of acid catalysts include both fuels (refinery applications) and chemicals production. The mechanisms by which acid catalysts produce these important high volume products are discussed in Secs. 2.3 and 2.4. The refinery applications are discussed first in Sec. 2.3.

2.3.1 *Catalytic Cracking*

An example of hydrocarbon cracking is the conversion of hexane to propane and propylene. In refinery operations, this reaction is referred to as *catalytic cracking* and is an important means by which small, fuel-grade hydrocarbons are produced from larger, heavier ones such as are found in crude oil. As was discussed in 2.2.1, it is believed that this process occurs on Lewis acid sites, which are so highly electron deficient that they are able to abstract a hydride from a paraffin (in this case, hexane) to form a 2° carbenium ion (2-hexyl) and the negatively charged Lewis site (Fig. 2.3.1). The carbenium ion undergoes β-scission to propylene and a 1° carbenium ion, which undergoes hydride transfer to form the more stable 2° carbenium ion. This 2° carbenium ion can then abstract a hydride from the original hexane molecule to again form the 2-hexyl carbenium ion. This last step is a "chain transfer" step, which is common in many catalytic reactions involving cationic and radical intermediates sych as polymerization reactions (Chapter 4).

When longer-chain paraffins are cracked catalytically, the same basic reactions occur, but the olefins formed can protonate to form new carbenium ions which can undergo the β-scission and isomerization reactions shown in Fig. 2.3.1.

The steps for catalytic cracking of a normal paraffin have been summarized (Ref. 5):

(1) Hydride abstraction on Lewis sites to form 2° carbenium ion.
(2) β-scission to form propylene or a higher olefin and a 1° carbenium ion.

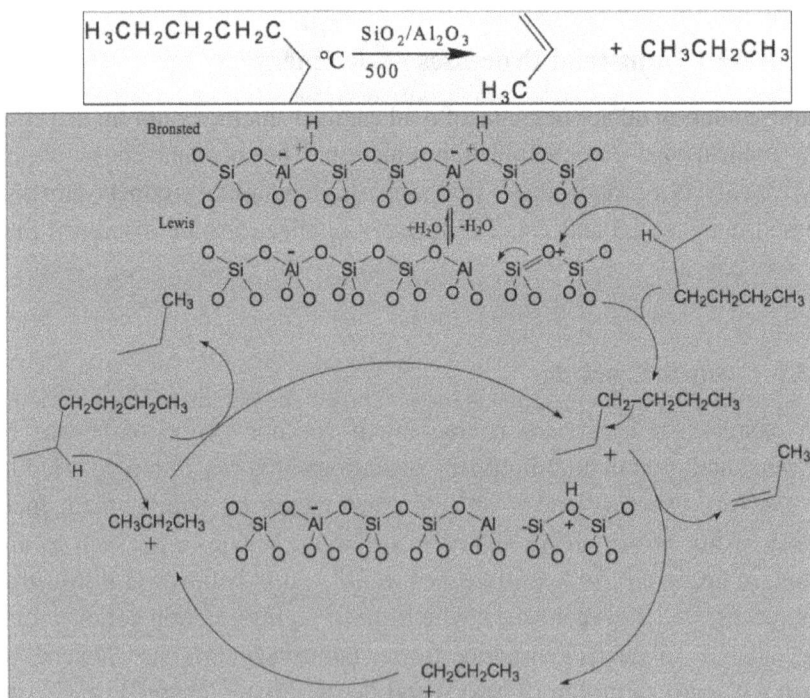

Fig. 2.3.1   Catalytic Cracking (Refs. 1, 5, 12)

(3) Isomerization of 1° to 2° carbenium ion.
(4) (<6 carbon atoms): Hydride abstraction by 2° carbenium ion to form paraffin and 2° carbenium ion, which reacts according to Step 1 or (if >6 carbon atoms), by Step 2.
(5) Higher olefins (>7 carbon atoms) from Step 2 protonate to form carbenium ions which react according to Steps 3 and 4 until they make propylene.

Zeolites are the catalysts of choice because their confined cavities prevent polymerization, which leads to deactivation by heavy molecules and coke formation (Secs. 2.2.2–2.2.4). Zeolites with metals can isomerize paraffins at low temperature to produce branched isomers desired for octane value (Sec. 5.3.5).

2.3.2 *Disproportionation of Aromatics*

The disproportionation of aromatics is the reaction which generates multialkylated (and non-alkylated) aromatics from monoalkylated ones. This is a common reaction in refining which is used to control the octane level of gasoline. When it occurs in the large pores of a zeolite such as ZSM-5 or mordenite, the relative amounts of the products which form reflect the ease with which they can diffuse through the opening in the supercage. This "shape selective" mechanism produces the less space-filling isomers (Sec. 2.2.4). Reactions within the ZSM-5 supercage give this catalyst a long life because it is not suspectable to coking. (Sec. 2.2.3)

The disproportionation reaction (Fig. 2.3.2) occurs by proton transfer to the mono-alkylated aromatic (toluene) in the first, rate-determining step on Bronsted acid sites. The protonated aromatic transfers an alkyl (methyl) cation to another molecule of toluene to produce benzene and the protonated dialkyl aromatic (xylene). The latter losses a proton to another molecule of toluene to produce xylene and the original protonated toluene cation. This is another example of a chain transfer mechanism.

In a chain transfer mechanism, the real catalytic species are the [aromatic]$^+$/[zeolite]$^-$ ion pair since this is the species that is regenerated. The protonated form of the zeolite, while it is usually referred to as the "catalyst," it is really in this case, the catalyst precursor or *initiator*. The zeolite structure, none the less, is involved as the catalytic species because

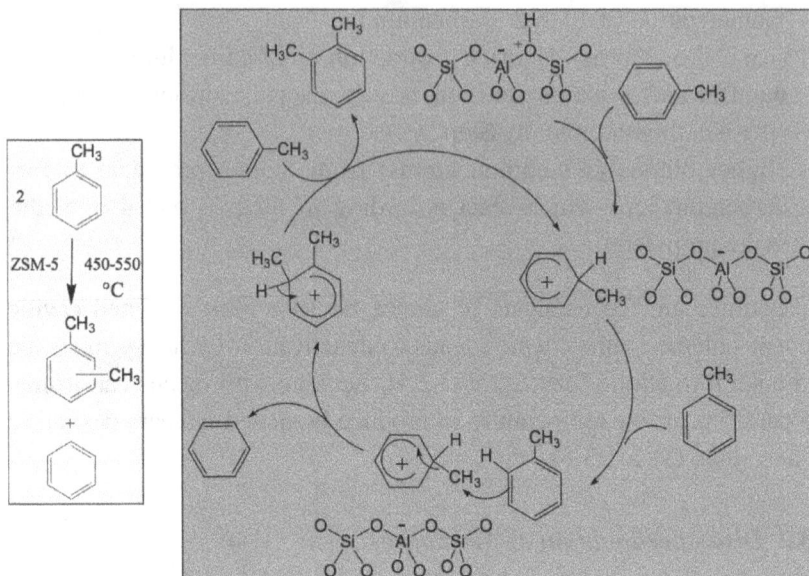

Fig. 2.3.2   Disproportionation of Aromatics (Refs. 5, 12)

the deprotonated zeolite anion serves as the counter ion for the cationic intermediates in the catalytic cycle. For chain transfer mechanisms, the "catalyst" that gets the chain started is referred to as an *initiator*. These are common in polymerization, where many of the reactions occur via a cationic, radical or anionic chain transfer mechanism (Chapter 4).

2.3.3   *Isobutane/isobutylene Alkylation (Octane)*

Isobutane/isobutylene alkylation (Fig. 2.3.3) is another important means by which octane value of gasoline is increased. This method is becoming increasingly important as more restrictions are placed on the levels of aromatics allowed in gasoline. In this reaction, a C_8 molecule — isooctane, a high octane liquid — is produced from two C_4s. Liquid acids such as HF or 100% H_2SO_4 are used as catalysts. While many attempts have been made to develop a heterogeneous solid acid system for this reaction, none have been able to compete with the catalytic performance and low cost of these liquid acids.

    In the first step, isobutylene is protonated to produce a *tert*-butyl cation, which reacts with another molecule of isobutylene to produce the

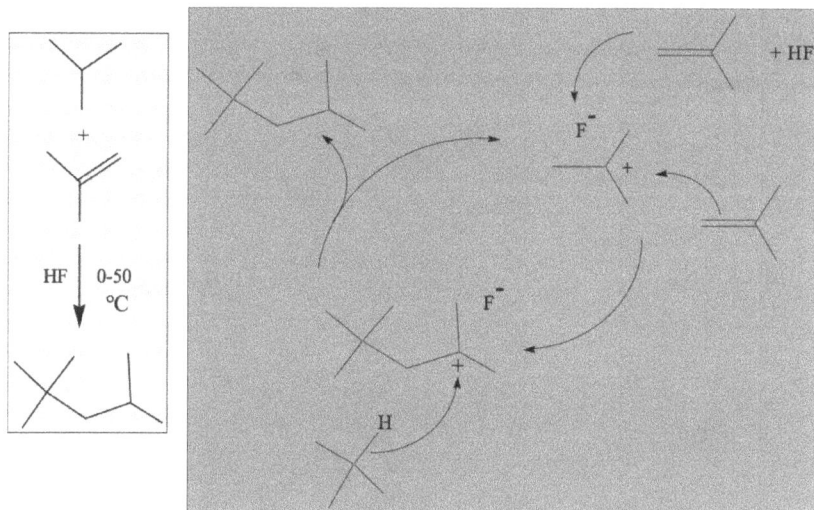

Fig. 2.3.3 Isobutane/Isobutylene Alkylation (Octane) (Refs. 1,5)

isooctyl cation. This cation then abstracts a hydride from isobutane to produce isooctane and the tert-butyl cation (hydrde abstraction by a carbenium
ion, as in catalytic cracking — Sec. 2.3.1). A critical feature of the process
is to favor hydride abstraction by isooctyl cation from isobutane, versus
reaction of isooctyl cation with isobutylene to form an olefinic C_8 dimer,
which, when repeated, eventually would lead to polymer (polyisobutylene).

Since the latter reaction (which does not involve isobutane at all)
competes favorably with the former based on thermodynamics, an excess
of isobutane must be used to produce the desired C_8 paraffin — isooctane.
However, in order for the process to be economic, the ratio of isobutane:
isobutylene should be kept at or below about 6:1. A critical means of
keeping this ratio in this range is the counterion, in this case F^-. This
favors hydride abstraction (from isobutane) leading to isooctane versus
olefin (isobutylene) addition, which leads to polymer. This is, once again,
an example of a chain transfer mechanism, where HF is the initiator and
the catalytic species is the $[tert\text{-butyl}]^+/F^-$ ion pair.

2.3.4 *Methyl–tert–butyl ether Synthesis*

Methyl–tert–butyl ether (MTBE) is an octane improver that had gained wide
acceptance, until the determination that it may leach into soil to produce

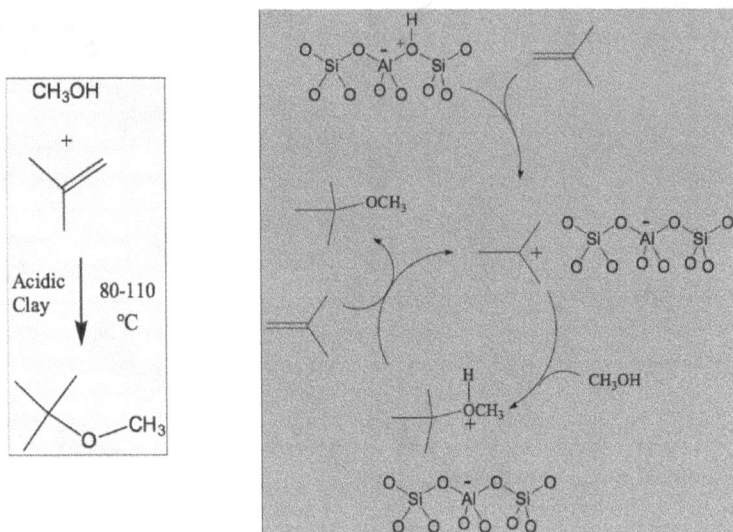

Fig. 2.3.4   MTBE Synthesis

unsafe levels in groundwater. While its use is limited, it does illustrate the application of a solid acid to convert an olefin and an alcohol to an ether.

The formation of MTBE over silica–alumina solid acid catalyst as shown in Fig. 2.3.4, is a straightforward protonation of isobutylene to form the tert-butyl cation (same initiation as for isobutylene/isobutane alkylation), followed by nucleophilic attack by methanol and proton transfer to isobutylene to regenerate the tert-butyl cation. Again, this is a chain transfer mechanism.

2.3.5 *Methanol-to-Gasoline (MTG)*

The Mobil methanol-to-gasoline process is a synthetic fuels process which is catalyzed by the zeolite ZSM-5. Many mechanisms have been proposed for this process, but it is generally understood that the first formed product is low temperature dehydration to diethyl ether, followed by further dehydration with C–C bond formation at higher temperature (400°C), as shown in Fig. 2.3.5.

Most mechanistic explanations refer to this as a two-step process in which methanol is first dehydrated to ether, and then a second higher

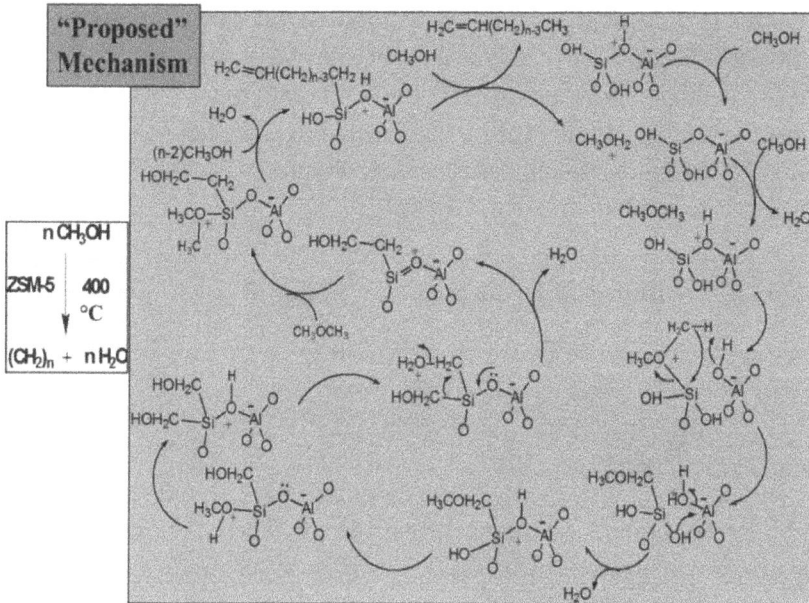

Fig. 2.3.5 MTG Mechanism (Refs. 1, 5)

temperature dehydration to produce a "CH_2" (methylene) group on the catalyst surface. These methylene groups the polymerize according to certain rules of chain transfer: chain termination statistics (same as for the Fischer–Tropsch gas to liquids process in Chapter 5), to form a distribution of hydrocarbons that is within the gasoline range.

The formation of gasoline range hydrocarbons versus higher (>C12) hydrocarbons over non-shape selective catalysts is an example of restricted transition state shape selectivity. This overall "mechanism" is shown in Eqs. 2.3.5:

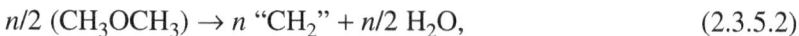

$$n\ CH_3OH \rightarrow n/2\ (CH_3OCH_3) + n\ H_2O, \tag{2.3.5.1}$$

$$n/2\ (CH_3OCH_3) \rightarrow n\ "CH_2" + n/2\ H_2O, \tag{2.3.5.2}$$

$$n\ "CH_2" \rightarrow (CH_2)_n\ n = 4\text{--}16\ \text{(gasoline range)}. \tag{2.3.5.3}$$

This "mechanism" does not really provide any details on how the C–O bonds in the ether are broken nor how the C–C bonds in the ether are broken nor how the C–C bonds in the final product are formed. An attempt to do this is illustrated in 2.3.5. The "proposed" mechanism shows how a Bronsted acid site might catalyze the ether formation and its conversion to a gasoline range hydrocarbon with an olefin end group. This is only one of many possible mechanisms, so the details of it are not so important as is its illustration of the importance of acidic protons at the Si–O site for C–C bond formation and dehydration.

2.4 Major Industrial Processes — Chemical

The processes discussed in the remaining sections are used for production of chemicals.

2.4.1 *Benzene Ethylation*

Ethylbenzene production from benzene and ethylene is the first step in the manufacture of styrene (by the SMPO — process, see Sec. 3.3.1). Styrene is a very large-scale monomer; about 14 million tons/yr are made for production of polystyrene, styrene butadiene rubber (mainly used in tires) and acrylonitrile-styrene-butadiene (ABS resins).

The production of ethylbenezne is another illustration of aromatic alkylation mechanism (2.1.11) on Bronsted acid sites of a zeolite, and is shown in Fig. 2.4.1. Again, the choice of the zeolite is based on the

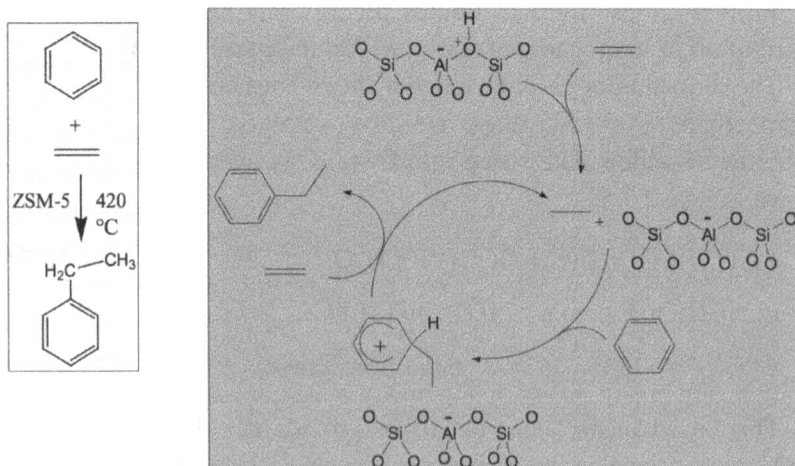

Fig. 2.4.1   Benzene Ethylation (Refs. 1, 12)

shape–selective feature of this catalyst, which is important in the prevention of deactivation by polymerization and coking, just as it is for catalytic cracking reactions (Sec. 2.3.1).

2.4.2 *Cumene*

Cumene (isopropylbenzene) is an intermediate used in the production of phenol and acetone (Sec. 3.3.9). If propylene is used instead of ethylene in the reaction mechanism of Sec. 2.4.1, the product cumene results. One of the catalysts used in this process is an example of an acid treated catalyst. The acidity of a support material is enhanced by its reaction with a Bronsted and/or a Lewis acid.

In this case, the support is silica, the Bronsted acid is phosphoric acid and the Lewis acid is BF_3. Several types of acid sites are formed by the reaction of hydroxl groups on the silica (Si–OH) with these acids. Condensation of SiOH groups with phosphoric acid produces a supported version of phosphoric acid ($-Si-O-PO_3H$), which is itself a very strong Bronsted acid (as shown in Fig. 2.4.2). This species can react with

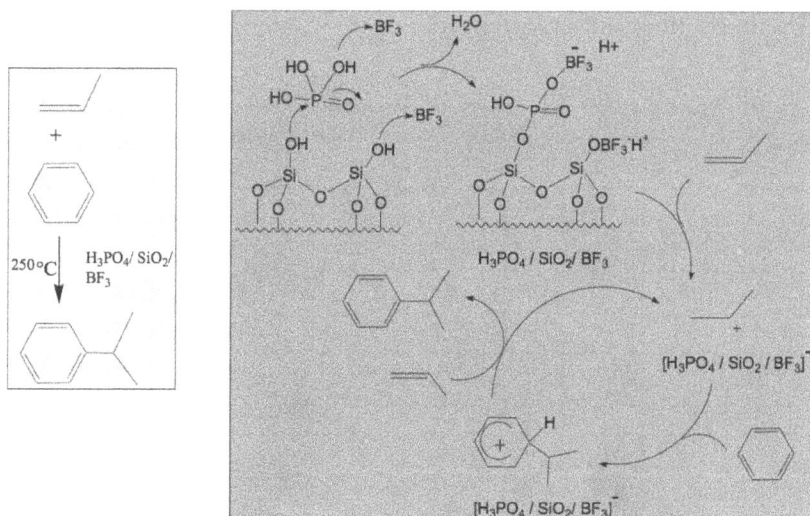

Fig. 2.4.2 Cumene Mechanism (Ref. 18)

BF$_3$ to produce an even stronger Bronsted acid which can be represented as [–Si–O–PO$_3$–BF$_3$]$^-$ H$^+$. Of course, the original Si–OH groups can also react with BF$_3$ to produce the strong Bronsted acid [–Si–O–BF$_3$]$^-$ H$^+$. All of these strong Bronsted acidic sites can initiate olefin protonation, the rate-determining step, which produces the isopropyl cation, the chain-carrying species in the catalytic cycle.

2.4.3 Beckman Isomerization

Caprolactam is a cyclic amide used in the production of Nylon 6. This is produced by the reaction of cyclohexanone oxime with a strong acid, usually sulfuric by the mechanism shown in Fig. 2.4.3.1 (non-catalytic). This reaction is sometimes referred to as acid "catalyzed" but in fact, the reaction is not catalytic, since a stoichiometric amount of sulfuric acid is needed to form the amide salt. The salt must then be neutralized in a second step with sodium hydroxide.

Several solid acids (represented as HX in Fig. 2.4.3.2 — Catalytic) have now been identified which can be used in catalytic amounts to produce caprolactam. This means that these solid acids can form an [amide]$^+$ X – salt, which is capable of proton transfer to cyclohexanone oxime to continue the catalytic cycle.

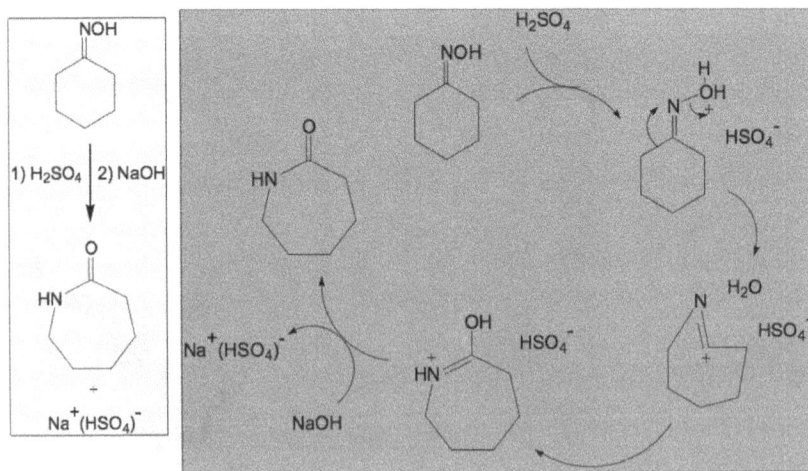

Fig. 2.4.3.1   Beckman Isomerization Mechanism (non-catalytic) (Ref. 18)

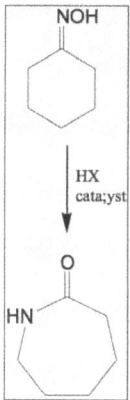

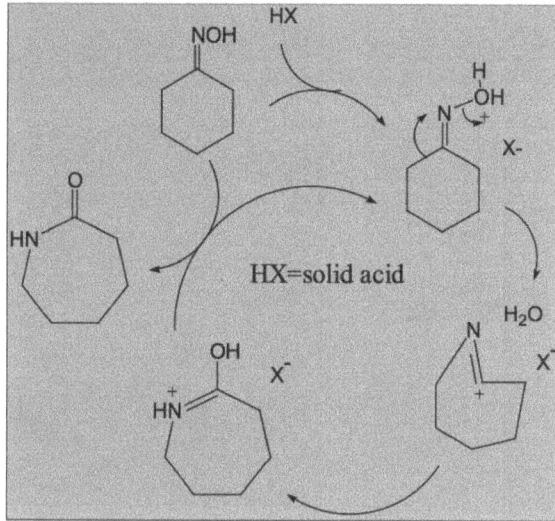

Fig. 2.4.3.2   Beckman Isomerization Mechanism (Catalytic)

2.4.4 *Esterification*

Ethyl acetate is a high volume chemical used as a solvent, mainly for paints. It can be continuously produced using a solid acid catalyst such as an acidic clay. A proposed mechanism involving an acidic silica–alumina clay site and an ethyl carbenium ion is shown in Fig. 2.4.4.

Clays can also be acidified with Bronsted and Lewis acids. Any number of acids can be used, the most common of which is sulfuric. This follows the same Si–OH condensation mechanism as was illustrated for the solid phosphoric acid catalyst shown in Fig. 2.4.2. In this case, the strongly acidic species ($-Si-O-SO_3H$) would be formed.

2.5 Trends in Acid Catalysis

Current research in solid acid catalysts indicates a number of trends for potential commercial application. A common theme among many investigations is the replacement of traditional Lewis ($AlCl_3$, BF_3) and mineral (H_2SO_4, HF) acids with solid acids, for several reasons. These echo the advantages of heterogeneous catalysts discussed in Chapter 1 (Sec. 1.7.4).

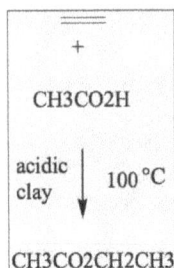

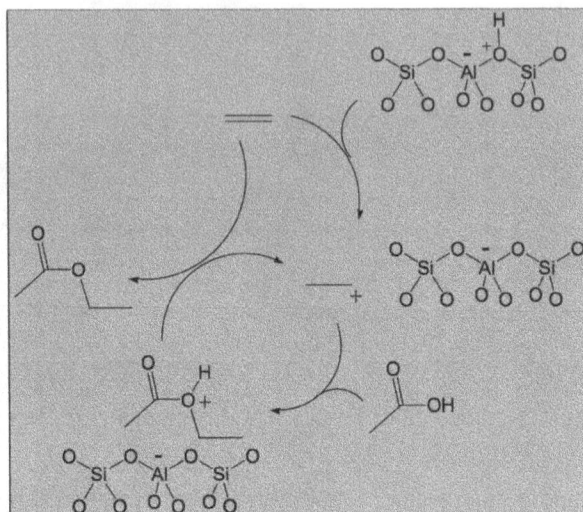

Fig. 2.4.4   Esterification

2.5 Trends in Acid Catalysis Conversion of Lewis and Mineral Acids to Solid Heterogeneous Catalysts

- Continuous Reactions
- Recyclable Catalysts
- Less Aqueous Waste
- Better Raw Material Utilization
- Examples (current catalyst to be replaced):
  - Isobutane/Isobutylene Alkyation (HF)
  - Beckman Isomerization to Caprolactam (H_2SO_4)

First, solid, heterogeneous acid catalysts are amenable to continuous processing and are generally stable and recyclable. Solid acid catalysts do not generally require a water-quench and are not water soluble. Therefore, aqueous waste is minimized.

In addition, solid acid catalysts can be confined to the reactor and therefore contact with product is minimized. This reduces acid-catalyzed decomposition or other post-reactor degradation of product, thereby improving selectivity and better utilization of raw material. Current examples of the use of solid acids for commercially significant products include refinery alkylation (Sec. 2.3.3) and Beckman isomerization (Sec. 2.4.3), as were discussed earlier in this chapter.

2.6 Problems

1. What active catalyst is formed when $AlCl_3$ is mixed with methanol (CH_3OH)? Is this a Bronsted or a Lewis acid?
2. What is the most common unselective product in acid-catalyzed reactions? Why is ZMS-5 used to reduce its formation?
3. Which silica–alumina catalyst would be better for cracking of an alkyl aromatic: (a) one calcined at 350°C or (b) one calcined at 750°C. Why?
4. For the following reactions of ortho-xylene over a mordenite or ZSM-5 catalyst: Why does isomerization predominate over disproportionation?

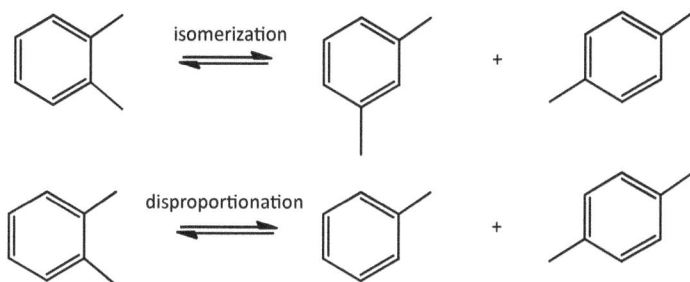

This is an example of what type of shape selectivity mechanism?
(Hint: Two molecules of xylene are involved in the disproportionation reaction.) In the isomerization reaction, which isomer would be expected to dominate?

5. When catalyst composed of Pt supported on alumina (Pt/Al_2O_3) is used to hydrogenate a 1:1 (molar) mixture of 1-hexene and 4,4-dimethyl 1-hexene, both are hydrogenated at equal rates:

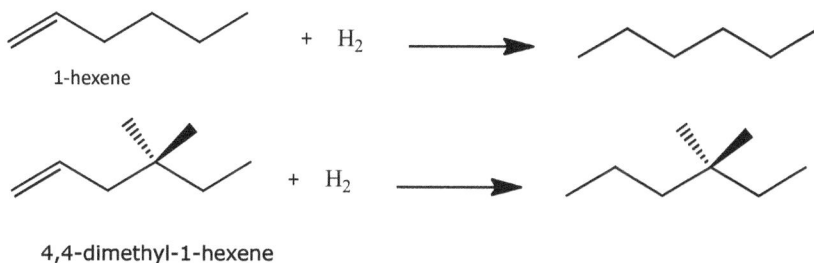

1-hexene

4,4-dimethyl-1-hexene

When Pt supported on ZSM-5 is used, the hydrogenation of 1-hexene occurs about 100 times faster. Explain.

2.7 Answers to Problems

1. $AlCl_3 + CH_3OH \rightarrow H + [AlCl3OR]-$. Bronsted acid.
2. What is the most common unselective product in acid catalyzed reactions? Coke or carbon. Why is ZMS-5 used to reduce its formation? The precursors to coke — polynuclear aromatic compounds — are very bulky and are formed by polymerization of aromatic species by processes that involve bulky transition states. The restricted ZSM-5 cavity makes these transition states energetically unfavorable thereby inhibiting coke formation. This is an example of *restricted transition state shape selectivity.*
3. Which silica–alumina catalyst would be better for cracking of an alkyl aromatic: (a) one calcined at 350°C or (b) one calcined at 750°C. Why? The high temperature (750°C) calcination generates the Lewis acid form, which is the active catalyst for cracking reactions. The one calcined at 350°C would make the Bronsted acid catalyst which catalyzes reactions involving activation by protontation, for example, aromatic alkylation reactions.
4. In the isomerization reaction, which isomer would be expected to dominate? Isomerization is a monomolecular reaction, i.e., one molecule is involved in the transition state, whereas disproportionation is a bimolecular reaction, i.e., two molecules are involved in the transition state, making the latter much more bulky and less favored to occur in the confines of the ZSM-5 zeolite cavity. This is an example of *restricted transition state shape selectivity* — i.e., the reaction with the less bulky transition state predominates.
5. The more bulky 4,4-dimethyl-1-hexene passes through the narrow opening of the ZSM-5 cavity with much greater difficulty than 1-hexene. Therefore, the former has much less access to the active Pt catalytic sites in the interior of the cavity. In fact, 1-hexene reacts about 100-times faster than 4,4-dimethyl-1-hexene over Pt/ZSM-5. This is an example of *reactant shape selectivity.*

3

Oxidation Catalysis

Oxidation of organic compounds is a highly exothermic process which produces CO_2 and water as the most thermodynamically-favored products. Selective oxidation, that is, the formation of partial oxidation products in preference to total oxidation (combustion) was not widely practiced commercially until the 1960s when selective oxidation catalysts were discovered. These catalysts contain elements which can cycle between several stable oxidation states, thus providing a low energy pathway to products of partial oxidation. Oxidation catalysts are essential for achieving high selectivity to partial oxidation products, particularly if air (or oxygen) is used as the stoichiometric oxidant.

In this chapter, oxidation catalysis is discussed, in terms of the constituent catalytic elements, reaction thermodynamics, the possible oxidation states of the reactants and products, and the basic mechanistic concepts by which selective oxidation catalysts operate.

The major parameters which allow oxidation catalysts to be designed and optimized for a particular process include the choice of metal, the isolation of active sites, the optimization of surface versus bulk composition, the use of promoter elements, and the particular synthetic method used. These important catalyst designed elements will be discussed along with examples of commercial catalyst design using these elements.

The chapter concludes with the presentation of representative examples of major industrial processes that use selective oxidation catalysis by reaction type: epoxidation, allylic oxidation, olefin/Wacker oxidation, paraffin oxidation and aromatic oxidation.

3.1 Concepts

The accessibility of multiple oxidation states provides a low energy pathway for electrons to flow from an organic reactant to a range of available electronic energy levels in the catalyst. The stoichiometric oxidant (i.e., the reactant in the chemical equation, which is reduced in the oxidation process) provides the oxidation potential necessary to maintain the catalyst in its most active, highest oxidation state. This is the basic mechanism by which catalytic oxidation of many organic compounds occurs.

3.1.1 *Oxidation Catalyst Periodic Table*

Elements which are commonly used for oxidation catalysts are multivalent, i.e., those which have several stable oxidation states (Table 3.1.1.1). These include the transition metals, the multivalent main group elements, and, of course, oxygen.

These transition metals and main group elements which best facilitate these redox processes — and are thus most common in oxidation

Table 3.1.1.1:   Oxidation Catalyst Periodic Table

1a	2a	3b	4b	5b	6b	7b	8			1b	2b	3a	4a	5a	6a	7a	0
H																	He
Li	Be											B	C	N	O	F	Ne
Na	Mg											Al	Si	P	S	Cl	Ar
K	Ca	Sc	Ti	V	Cr	Mn	Fe	Co	Ni	Cu	Zn	Ga	Ge	As	Se	Br	Kr
Rb	Sr	Y	Zr	Nb	Mo	Te	Ru	Rh	Pd	Ag	Cd	In	Sn	Sb	Te	I	Xe
Cs	Ba	La*	Hf	Ta	W	Re	Os	Ir	Pt	Au	Hg	Tl	Pb	Bi	Po	At	Rn
Fr	Ra	Ac**															

*Lanthanides	Ce	Pr	Nd	Pm	Sm	Eu	Gd	Tb	Dy	Ho	Er	Tm	Y	Lu
**Actinides	Th	Pa	U	Np	Pu	Am	Cm	Bk	Cf	Es	Fm	Md	No	Lw

Table 3.1.1.2: Oxidation Catalyst View of the Periodic Table

1a	2a	3b	4b	5b	6b	7b		8			1b	2b	3a	4a				0
H																O		He
Li	Be												B	C				Ne
V	**Cr**	**Mn**	**Fe**	**Co**	**Ni**	**Cu**							Al	Si		**S**		Ar
Nb	**Mo**	**Te**	**Ru**	**Rh**	**Pd**	**Ag**							Ga	Ge	**Sn**	**Sb**	**Te**	Kr
	Fr	Ra	Ac**														**Bi**	
*Lanthanides	Ce	Pr	Nd	Pm	Sm	Eu	Gd	Tb	Dy	Ho	Er	Tm						
**Actinides	Th	Pa	U	Np	Pu	Am	Cm	Bk	Cf	Es	Fm	Md	No	Lw				

catalysts — are shown in the enlarged view in Table 3.1.1.2 (along with O and S).

3.1.2 *Oxidation Thermodynamics*

The large driving force for the formation of carbon oxides compared to selective products (Table 3.1.3.1) is indicated by the free energies of reaction for total oxidation of propylene (−464 kcal/mole) versus formation of selective products acrylic acid (−131 kcal/mole) and acrolein (−81 kcal/mole.) The function of the selective oxidation catalyst is to provide a lower energy path for selective products, so that these are formed faster than CO_x, even though the free energy for total combusion is much greater.

3.1.3 *Recognizing Organic Oxidation States*

In the study of catalytic oxidation reactions, it is important to first be able to recognize the relative oxidation states of organic molecules (relative molecular oxidation states). These relative molecular oxidation states

3.1.3 Recognizing Organic Oxidation States

• Pertains to the overall "molecular" oxidation states of organic compounds
• Used to determine if an oxidation or reduction reaction has occurred

Table 3.1.3.1:   Oxidation Thermodynamics (Ref. 19)

Reaction	ΔG, kcal/mole
$C_3H_6 + 9/2O_2 \rightarrow 3CO_2 + 3H_2O$	-464
$C_3H_6 + 3O_2 \rightarrow 3CO_2 + 3H_2O$	-305
$C_3H_6 + 3/2O_2 \rightarrow CH_2 = CH-CO_2H + H_2O$	-131
$C_3H_6 + O_2 \rightarrow CH_2 = CH-CHO + H_2O$	-81

- Methodology is "virtual" transformations (not actual chemical reactions)
- Oxidation: If the virtual transformation involves formal loss of H_2; 2e⁻ Ox. for each H_2
- Reduction: If the virtual transformation involves formal loss of H_2; 2e⁻ Ox. for each H_2
- No Red/Ox: If the virtual transformation involves loss or addition of H_2O or no formal loss or addition of H_2

will determine if a given reaction is an oxidation (or a reduction) reaction, or if it is redox neutral (i.e., no oxidation or reduction occurs.)

The methodology used to determine the relative oxidation states of two organic molecules is to identify if a "virtual" transformation of one molecule to the other involves the addition or elimination of a "H_2" or "H_2O". An oxidation reaction involves the formal loss of "H_2" (2e⁻ for each "H_2"), and a reduction reaction, the corresponding addition of "H_2". The addition or elimination of water is a redox neutral process.

Relative Oxidation State Above Paraffin

= No. of Carbon–Carbon Multiple Bonds
+ No. of Carbon–Hetero-atom Bonds
+ No. of Carbon Rings.   (Eq. 3.1.3.1)

Based on this methodology, the relative oxidation state above the paraffin state can be calculated by Eq. 3.1.3.2:

Relative oxidation state (above paraffin) = ROS

$= (\Sigma\text{ C–C multiple bonds}) + (\Sigma\text{ C–X bonds}) + (\Sigma\text{ C–rings})$, where
$\Sigma\text{ C–C multiple bonds} = C = C + 2* C \equiv C$

Σ C–X bonds$=$(C–X)$+2*$(C$=$X)$+3*$(C$=$N)

    X$=$a heteroatom (e.g., O, N, S), and where

    Σ C–rings$=$number of rings composed only of C atoms.

For example, a paraffin has no C–C multiple bonds, no C–X bonds and no rings, so it has an ROS of 0. A molecule with an ROS$=n$, can also be referred to as at the "P$+$n" oxidation state. A cyclic paraffin would have an ROS of 1 or P$+$1. Examples of relative oxidation states for other organic molecules are shown in Table 3.1.3.2.

Examples of the oxidation states of organic molecules with C, H and O are shown in Table 3.1.3.2, and are organized by three functional groupings: hydrocarbons (molecules with C and H only), hydroxy-keto (molecules with C–O single bonds) and ketoacids (molecules with carbonyls (C$=$O).

Examples of the most common 6-carbon organic molecules in these categories are shown. For example, benzene has an oxidation state of P$+$4 because it has 3 C$=$C bonds and 1 ring. Cyclohexene oxide has an oxidation state of P$+$3 (2 C–O bonds and 1 ring). Therefore, conversion

Table 3.1.3.2: Organic Oxidation States

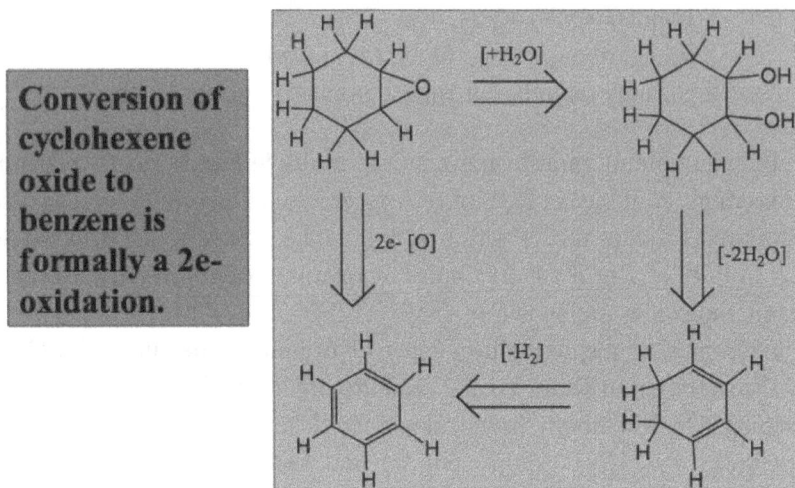

Fig 3.1.3.1 Organic Oxidation States — "Virtual" Transformation Example

of cyclohexene oxide to benzene is a 2-electron oxidation and would therefore require some oxidant.

In contrast, cyclohexane 1,2-diol can be thought of as resulting from addition of H_2O to cyclohexene oxide, and therefore, the two molecules are of the same oxidation state (i.e., both are P+3). The virtual conversion of cyclohexene oxide to benzene is shown in Fig. 3.1.3.1.

In the "virtual" conversion of cyclohexene oxide to benzene, the first "step" is the addition of water, which produces cyclohexane 1,2-diol, which upon loss of water produces cyclohexadiene. Because all of these "virtual" transformations involved only the addition or loss of water, all three of these molecules are at the same oxidation state (P+3). However, to convert cyclohexadiene to benzene requires the "virtual" loss of 1 molecule of H_2. Therefore, the overall conversion of cyclohexene oxide to benzene is formally a 2-electron oxidation. Conversion of either of the other two molecules to benzene is also a 2-electron oxidation.

Some other examples from Table 3.1.3.2 are highlighted in Fig. 3.1.3.2. The four Molecules at the P+2 level can all be interconverted by "virtual" reactions involving only the addition or elimination of water.

The following are all at the P + 2Level:

The following are all at the P + 7 Level:

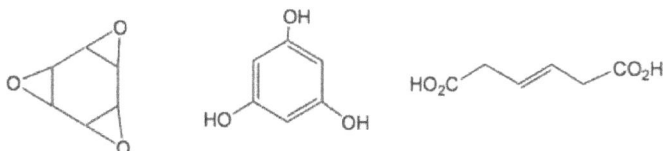

Fig. 3.1.3.2 Organic Oxidation States — Examples of Same Oxidation State

• 2 e- Oxidation:

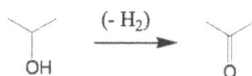

• 2 e- Reduction:

$$(+ 2 \text{ H}_2)$$
$$\text{CO} \longrightarrow \text{CH}_3\text{OH}$$

• No Rex/Ox

$$(+\text{H2O})$$
$$\text{HC}{\equiv}\text{CH} \longrightarrow \text{CH}_3\text{CHO}$$

Fig 3.1.3.3 Organic Oxidation States — "Virtual" Transformations

This "virtual" reaction methodology is only valid for organic molecules with the same number of carbon atoms. For reactions involving formation or breaking of C–C bonds, the redox nature of the reaction must be determined from the stoichiometry of the reaction.

In Fig. 3.1.3.3 are some examples of reactions that involve oxidation (alcohol to ketone), reduction, CO to methanol, and no redox (hydration of acetylene). The "virtual" loss and addition of H$_2$, and addition of water, respectively can be used to determine the redox nature of these processes.

3.1.4 *Oxidation Catalysis Concepts — Homogeneous*

The mechanisms for catalytic oxidation can be divided into homogenous (Fig. 3.1.4) and heterogeneous (Fig. 3.1.5) categories. There are two basic types of homogeneous oxidation reaction mechanisms: molecular and free radical. In the molecular route, a soluble metal species (shown in Fig. 3.1.4 as a metal oxide) reacts with an active oxygen species such as a peroxide to form metal peroxy species. This peroxo species reacts with a hydrocarbon (RH) to form selective oxidation products and regenerate the metal oxide. In the free radical route, an oxidized metallic species abstracts a hydrogen atom from a hydrocarbon (RH) to form an organic radical R• and a reduced metal species. The organic radical reacts with oxygen, and subsequently with the reduced metal to form selective oxidation products and the oxidized metal species (as in 3.1.7.2 — Homogeneous Free Radical Mechanism).

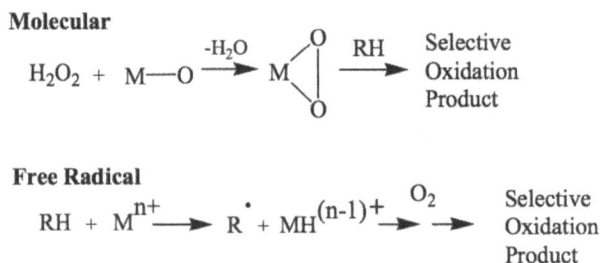

Molecular

$$H_2O_2 + M\!-\!O \xrightarrow{-H_2O} M\!\!\underset{O}{\overset{O}{\big<}} \xrightarrow{RH} \text{Selective Oxidation Product}$$

Free Radical

$$RH + M^{n+} \longrightarrow R^{\bullet} + MH^{(n-1)+} \xrightarrow{O_2} \longrightarrow \text{Selective Oxidation Product}$$

Fig. 3.1.4 Homogeneous Oxidation Catalysis Concepts

3.1.5 *Oxidation Catalysis Concepts — Heterogeneous*

Heterogeneous oxidation catalysis can also occur by two major mechanistic types: surface and solid state (Fig. 3.1.5). In the surface mechanism, a metal surface activates oxygen by forming a surface oxide species, which then reacts with hydrocarbon to form selective oxidation products and the original metal surface. In this mechanism, only oxygen atoms which are on the surface, or only a few layers down, participate in the reaction.

In contrast, many layers of oxygen atoms in the bulk participate in the reaction in the solid state (also know as the "Mars–van Krevelen") mechanism (Fig. 3.1.5, bottom left). In this mechanism, a hydrocarbon is

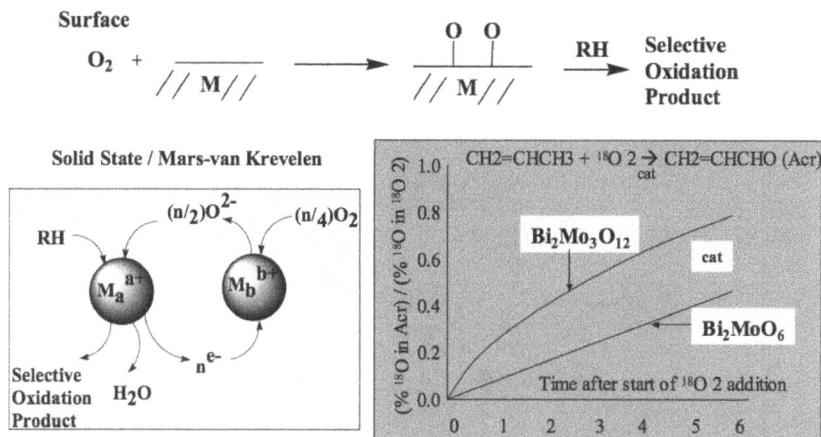

Surface

$$O_2 + \underset{//\ M\ //}{\underline{\hspace{2cm}}} \longrightarrow \underset{//\ M\ //}{\overset{\overset{\displaystyle O \quad O}{\|\quad\ \|}}{\underline{\hspace{2cm}}}} \xrightarrow{RH} \text{Selective Oxidation Product}$$

Solid State / Mars-van Krevelen

CH2=CHCH3 + $^{18}O_2 \xrightarrow{cat}$ CH2=CHCHO (Acr)

Fig. 3.1.5   **Heterogeneous Oxidation Catalysis Concepts** (Ref. 19, reproduced with permission)

oxidized by donation of the bulk oxygen atoms of the metal oxide lattice at site M_a, to form selective oxidation products, water and a reduced-metal oxide. Electrons for site M_a then flow to a second metal oxide site M_b, where oxygen is reduced to oxide ions, which then flow back to M_a to refill the oxygen vacancies. Thus, the Mars–van Krevelen-type catalysts can be thought of as a surface oxidation site (M_a) and an oxygen reduction site, which are electronically and structurally coupled by the shared oxide ions of the bulk solid state lattice. These oxide ions are depleted during the oxidation step and are replenished during the reoxidation step.

3.1.6 *Elementary Reaction Steps*

Two important mechanistic steps in oxidation catalysis are illustrated in Fig. 3.1.6. The mechanism for the oxidation step for heterogeneous oxidation by metal oxides (Mars–van Krevelen-type catalysts) involves the abstraction of hydrogen by a M=O double bond, with formation of a carbonyl (2e⁻ oxidation of the organic species and 2e⁻ reduction of the metal (Fig. 3.1.6, Eq. 3.1.6.1). The H-abstraction and reduction are shown as occuring on the same metal, but these may actually happen on two separate metals, with each one undergoing a 1–e reduction.

Oxidation by Metals – O_2:

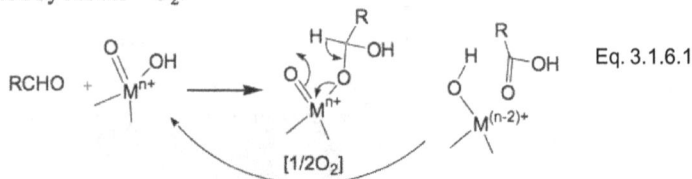

Eq. 3.1.6.1

Oxidation by Metals - Peroxide:

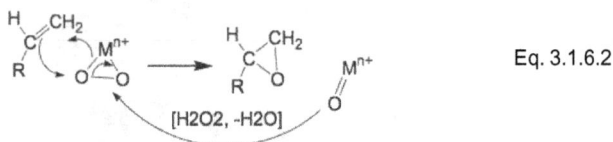

Eq. 3.1.6.2

Fig. 3.1.6   Elementary Reaction Steps

For homogeneous (molecular) oxidation, as illustrated in Fig. 3.1.6, Eq. 3.1.6.2 for the epoxidation of an olefin by hydrogen peroxide, the peroxy species is the active oxidant, which undergoes breaking of the weak O–O bond and formation of the stronger M=O bond and 2 C–O bonds, providing a strong driving force for the reaction.

3.1.7 *Homogeneous Oxidation Catalysis*

Homogeneous oxidation catalysis mechanism can be divided into two classes: molecular (Fig. 3.1.7.1) and free radical (Fig. 3.1.7.2).

3.1.7.1 **Homogeneous molecular**

The mechanism for the olefin epoxidation using an alkyl peroxide and a soluble Mo catalyst is shown in Fig. 3.1.7.1. In this case, a metal (Mo) alkyl peroxy species forms with loss of an –OR ligand. This species undergoes weak O–O bond breakage with formation of the epoxide C–O bonds and loss of ligand (L) in the olefin oxidation step. The L–Mo bond reforms in the product dissociation step.

Another important homogeneous catalytic oxidation reaction is the Wacker oxidation for conversion of an olefin into an aldehyde (Fig. 3.1.7.1.2). This reaction is commercially used for the production of

**Olefin
Epoxidation**

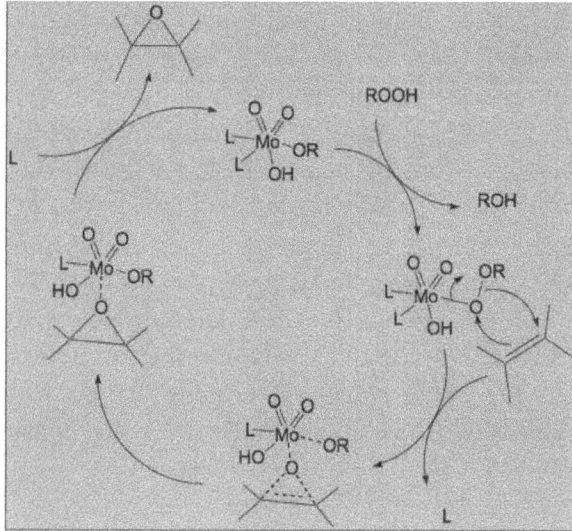

Fig. 3.1.7.1.1   **Homogeneous Molecular Oxidation: Epoxidation** (Ref. 6, reproduced with permission)

**Wacker
Oxidation**

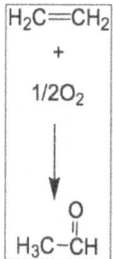

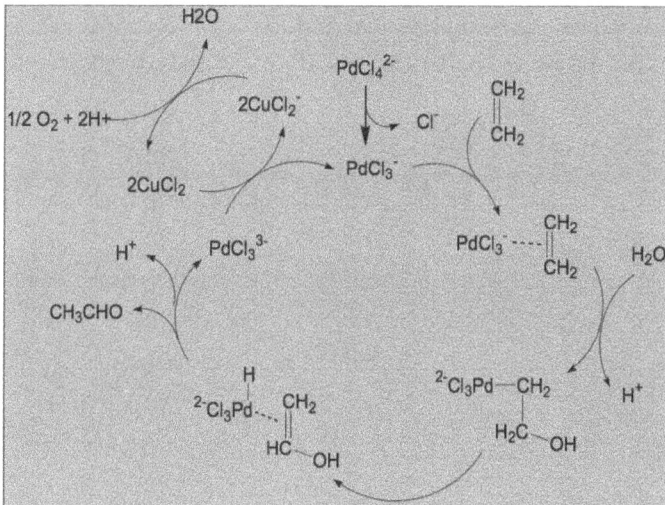

Fig. 3.1.7.1.2   **Homogeneous Molecular Oxidation — Wacker Oxidation** (Ref. 6, reproduced with permission)

acetaldehyde from ethylene using O_2 (from air) as the stoichiometric oxidant. In this reaction, a Pd/Cu metal oxidation couple is used. Pd^{2+} oxidizes the ethylene, and the resulting Pd^0 species is reoxidized by Cu^{2+}. The Cu^+ formed in the latter process is then reoxidized by O_2. In the product-forming sequence, the Pd^{2+} species activates olefin to addition of water, and undergoes β-elimination to form acetaldehyde enol, which desorbs and rapidly tautomerizes (H-transfer from O to C) to form product acetaldehyde. In the process, Pd^0 and 2 protons are formed, and one water is consumed. In the reoxidation step (mediated by the Cu^{2+}/Cu^+ couple), $\frac{1}{2} O_2$ and the 2 protons are consumed, and one water is formed. (Water and protons are not involved in the stoichiometry because equal amounts are formed and consumed in each catalytic cycle).

3.1.7.2 Homogeneous free radical

Homogeneous catalytic oxidations can also occur by a free radical mechanism. An example of this is the air oxidation of toluene to benzoic acid using a Co^{3+} catalyst (Fig. 3.1.7.2). Co^{3+} acts as a 1−e oxidant, producing the benzyl radical and Co^{2+}. The benzyl radical reacts with O_2 to form a benzyl peroxide radical. Co^{2+} catalyzes the decomposition of the

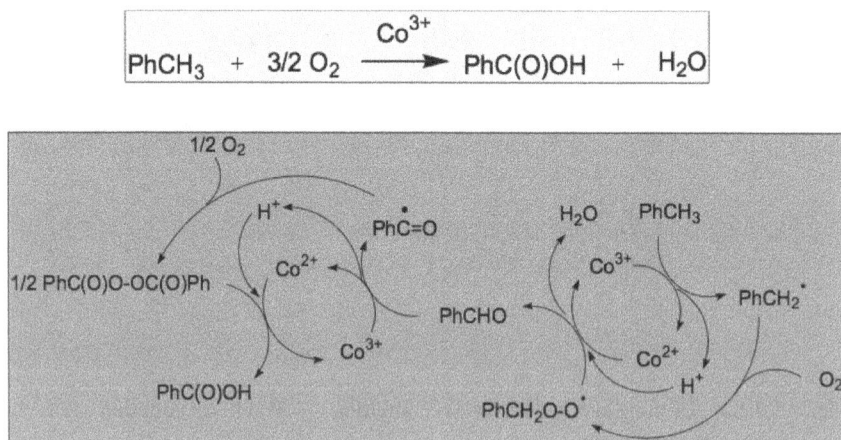

Fig. 3.1.7.2 Toluene to Benzoic Acid (Ref. 5)

benzyl peroxide to form benzaldehyde and water and regenerate Co^{3+}. The proton from the initial benzyl radical formation is consumed in this benzaldehyde-and-water-forming step.

3.1.8 *Hetereogeous Oxidation*

Heterogeneous Oxidation Catalysis can be divided into two classes: Surface (Fig. 3.8.1.1) and Bulk (Fig. 3.8.1.2).

3.1.8.1 Hetereogeous surface oxidation

As we have seen, heterogeneous oxidations occur by reaction of hydrocarbon at the surface of the catalyst, but can also involve various levels of participation of the bulk oxide lattice of the catalyst. On one extreme is surface oxidation with little or no involvement of bulk oxide (surface oxidation mechanism), and on the other, participation of many layers of the bulk (Mars–van Krevelen-type mechanism).

The former is illustrated in Fig. 3.1.8.1 by the epoxidation of ethylene by O_2 over a silver catalyst. In the presence of oxygen, silver is oxidized

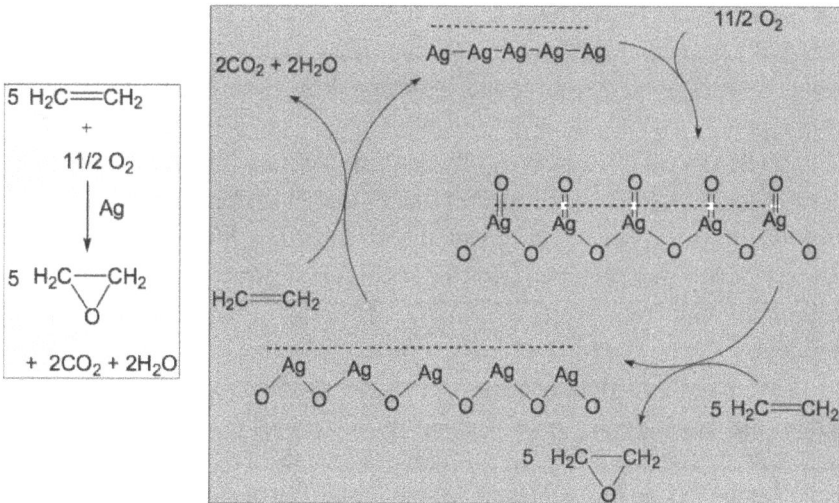

Fig. 3.1.8.1   Heterogeneous Surface Mechanism: Ethylene Oxide (Ref. 1)

to form an oxide with two types of oxygens: Ag–surface bonds which have double-bond character (Ag=O) and Ag–O single-bonded bridging oxygens. The surface Ag=O species is thought to be the selective component and bridging oxygen the unselective (CO_2-forming) sites. As shown in Fig. 3.1.8.1, for every 11/2 O_2 (or 11 O atoms) which react, 5 oxygen atoms form selective Ag=O sites, which result in the formation of 5 moles of ethylene oxide from 5 moles of ethylene. The other 6 oxygen atoms form unselective bridging oxygens, which react with one mole of ethylene to form 2 moles of CO_2 and one mole of water. Thus, for every 6 moles of ethylene, 5 form ethylene oxide (selective product) and 1 forms CO_2 (unselective product). This means that the maximum selectivity for the reaction by this mechanism should be 5/6 or 83%. This is in fact, the maximum selectivity that has been found for the (unpromoted) Ag/O_2 system. Selectivities greater than 83% can be obtained by the of use of Cl as a promoter (3.2.4, 3.2.5).

3.1.8.2 Heterogeneous bulk (Mars–van krevelen) oxidation

A representative example of a heterogeneous catalytic oxidation which proceeds via the Mars–van Krevelen-type mechanism is shown in 3.1.8.2: the selective oxidation of propylene to acrolein over bismuth molybdate catalysts such as $Bi_2Mo_3O_{12}$. In this mechanism, the surface site is composed of a bismuth H-abstraction component, and an oxygen inserting Mo component. The allyl radical species formed by H–abstraction on the Bi component is the rate-limiting step, which results in a $1e^-$ reduction of Bi (3+) to Bi (2+). This adsorbed allyl radical reacts with a Mo=O double bond to form a C–O bonded species (the O-insertion step). The O-insertion step results in a $1e^-$ reduction of the O–inserting Mo(6+) to Mo(5+). A second Mo atom then performs the second H-abstraction and undergoes a $2e^-$ reduction from Mo(6+) to Mo(4+) forming of the selective product acrolein.

The overall oxidation of propylene to acrolein is a $4e^-$ oxidation, where the corresponding $4e^-$ reduction of catalyst is shared among an initial H-abstracting Bi atom ($1e^-$ reduction), the O-inserting Mo atom ($1e^-$ reduction) and a second H-abstracting Mo atom ($2e^-$ reduction). These reduced metal centers produce O-vacancies in the lattice near the surface, which are rapidly replenished by bulk O with participation of

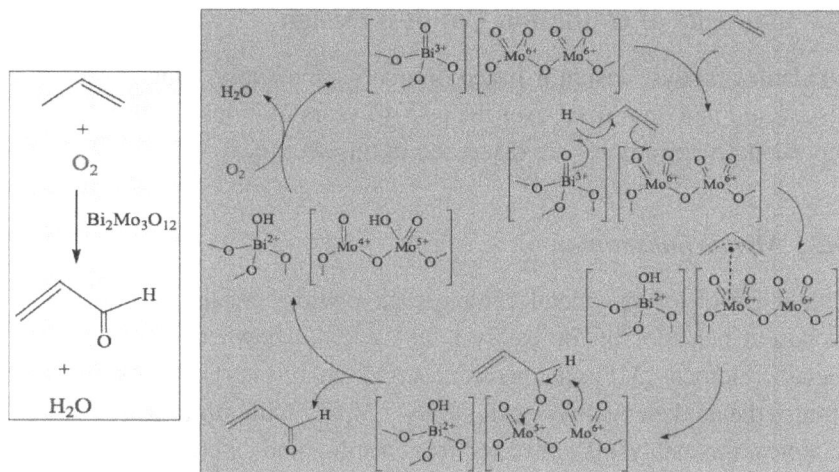

Fig. 3.1.8.2 Heterogeneous Bulk (Mars–van Krevelen) Mechanism: Oxidation of Propyene to Acrolein (Reproduced from Ref. 7 with permission)

Table 3.1.8.2: Mechanisms — Mars–van Krevelen Oxidation Catalysts (Ref. 20, reproduced with permission)

Catalyst	αH-Abstraction	O Insertion/2nd H-Abstraction	Redox Promoter
$Bi_2Mo_nO_{3+3n}$	$Bi^{3+ or 5+}$	Mo^{6+}	Ce^{3+}/Ce^{4+} or Fe^{2+}/Fe^{3+}
Te_2MoO_7	Te^{4+}	Mo^{6+}	Ce^{3+}/Ce^{4+}
$Fe_xSb_yO_Z$	Sb^{3+}	Sb^{5+}	Fe^{2+}/Fe^{3+}
USb_3O_{10}	U^{5+}	Sb^{5+}	U^{5+}/U^{6+}
$Fe_aSe_bTe_cO_x$	Se^{4+}	Te^{6+}	Fe^{2+}/Fe^{3+}

many layers in this process. Oxygen from the feed air reoxidizes the bulk, according to the general Mars–van Krevelen mechanism on 3.1.5. In this case the above active sites are "M_a" and the oxygen dissociation/reoxidation sites ("M_b") are believed to be associated with a Bi- rich site. Redox promoters such as Fe or Ce can be added to the catalyst to facilitate this reoxidation process.

The major catalytic components of H-abstraction, O-insertion/2nd H-abstraction and redox promotion are common to a number of solid oxidation catalyst systems. Several of these with the identification of which elements perform these functions are shown in Table 3.1.8.2.

3.2 Elements of Oxidation Catalyst Design

Oxidation catalyst design is based on many parameters, which have been discovered and optimized over the past 40 years of industrial use. Several representative examples are discussed in this section.

3.2.1 *Metals Substitution*

The selection of the metal center is an obvious means of controlling oxidation behavior of the catalyst. In the case shown in Fig. 3.2.1, the metallic element (M) in the heteropolyacid catalyst is shown. As one substitutes the acid element W in $H_{3+x}PW_{12-x}V_xO_{40}$ with more easily reduced V atoms, the catalyst is converted from a solid acid catalyst, which forms predominantly the dehydration ether product to an oxidation catalyst, which forms predominantly the oxidation product formaldehyde (CH_2O). This is illustrated in the graph in Fig. 3.2.1, which shows that the ratio of formaldehyde to dimethyl ether increases with increasing number of V atoms added to the catalyst.

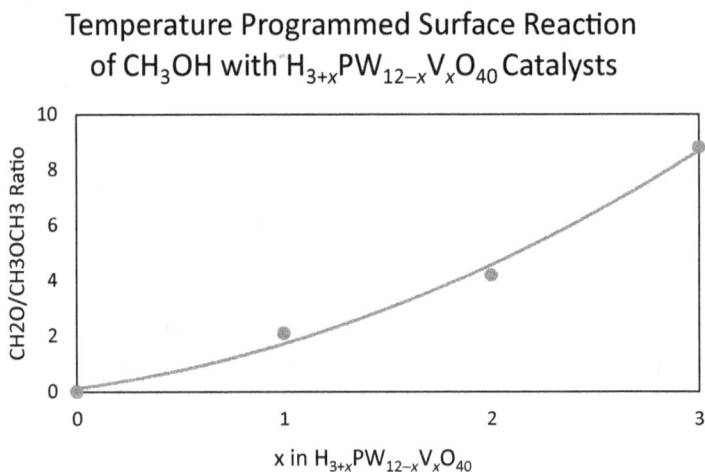

Fig. 3.2.1 **Temperature Programmed Surface Reaction of CH₃OH with H₃₊ₓ PW₁₂₋ₓVₓO₄₀ Catalysts** (Ref. 21)

3.2.2 *Active Site Isolation*

Active site isolation is another key concept for designing selective oxidation catalysts. A fully oxidized solid oxidant like Cu_2O_3 is very active, but unselective for the oxidation of an olefin such as propylene. However, the selectivity can be improved if the oxide is partially reduced using a reductant such as H_2 or an olefin. This phenomena is represented schematically in Fig. 3.2.2.1 by a hypothetical grid of oxidized (black circles) and reduced (open circles) sites. As the oxidized grid is reduced, the average number of oxidized sites which are adjacent to one another (a cluster) can be calculated from simple statistics (called a *Monte Carlo calculation*). The frequency distribution of the number of oxidized sites in such a continuous "cluster" as a function of % reduction is plotted in the upper right chart of Fig. 3.2.2.1, with the average number of sites/cluster in the boxes by each corresponding curve of the same color.

The actual reaction of propylene with air over CuO_x catalysts at various states of reduction is shown in the lower right chart. The yield of acrolein as a function of % reduction indicates that clusters with >5 oxidized sites per cluster are unselective; that is, they form mainly carbon oxides (CO and CO_2). Sites with much fewer than 5 oxidized sites are very inactive and produce little of any product. The maximum yield of selective product acrolein is produced at about 30% reduction, which corresponds to an average cluster of 5. This cluster size corresponds to a grid that is halfway between a and b, consistent with the importance of active (oxidized) site isolation for a selective oxidation catalyst.

While partial reduction does result in selectivity increase, it is not a very practical method for making a selective catalyst for industrial use. It is very difficult to maintain a given level of catalyst reduction in the presence of oxygen and hydrocarbon. A better approach is through site isolation by chemical means, as illustrated for the $USbO_x$ oxidation catalyst system in Fig. 3.2.2.2. (The actual use of U in a large-scale catalyst is, of course, not practical, but this example does serve to introduce the chemical site isolation concept. Examples of more practical industrial catalyst system will follow.)

In the $USbO_x$ catalyst system, oxygens associated with U (5+) atoms are the active oxidation sites, analogous to the oxidized sites in the grid

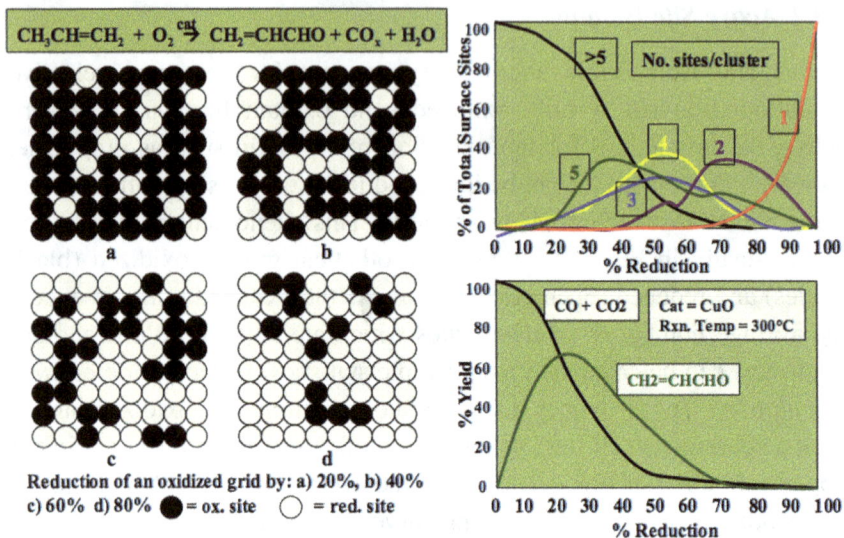

Fig. 3.2.2.1   **Active Site Isolation** (Ref. 19, reproduced with permission)

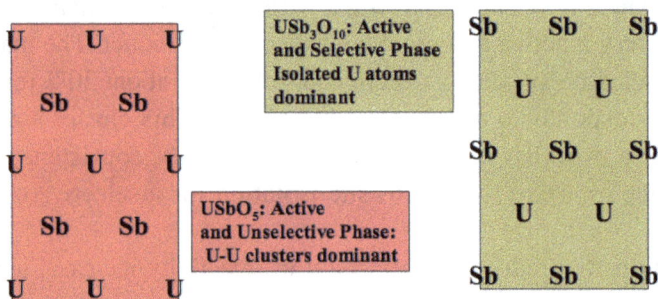

Fig. 3.2.2.2   **Active Site Isolation: Phases Present in USbO$_x$ Selective Oxidation Catalysts** (Ref. 19, reproduced with permission)

shown previously, while oxygens associated with Sb (5+) are less active, analogous to the reduced sites in this grid. When a U/Sb mixed metal oxide catalyst is made, two stable crystallographic phases form: USb$_3$O$_{10}$ and USbO$_5$. Cross-sections of these structures through the plane of U and Sb atoms (Fig. 3.2.2.2) show that the active oxidation centers are isolated in the USb_3O_{10} phase, while the $USbO_5$ phase has more clusters of adjacent U atoms. In fact, the former is the more selective phase consistent with isolation of active U sites by the less active Sb atoms. Since all the atoms in the USb_3O_{10} phase are in their highest oxidation state, and it is stable under the reaction conditions, chemical site isolation is a practical means to make a selective oxidation catalyst.

3.2.3 *Surface Versus Bulk Composition*

As mentioned previously, the surface of a heterogeneous oxidation catalyst can be compositionally very different from the bulk stoichiometry. This effect can be used to produce catalysts which are selective across a range of bulk compositions. This is an important aspect of catalyst life, since catalyst bulk compositions can change over time as elements are lost through phase changes and physical loss of catalyst (by a process known as *attrition*). A catalyst which can maintain its activity even after these bulk compositional changes ocur will be more robust and longer-lived.

This effect is illustrated by the $Bi_2O_3 \bullet nMoO3$ system, a common selective oxidation catalyst of industrial importance. The three points in the chart on the left of Fig. 3.2.3 correspond to the three major phases of catalytic significance: $n = 1, 2$ and 3: Bi_2MoO_6, $Bi_2Mo_2O_9$ and $Bi_2Mo_3O_{12}$. Each of these phases lies on the line with slope of one on a plot of surface

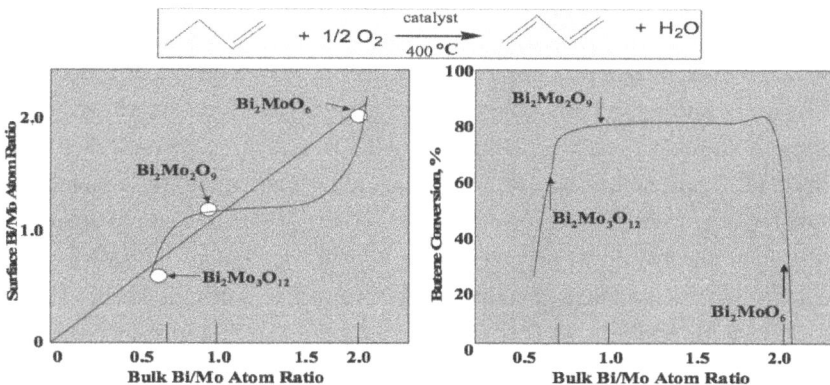

Fig. 3.2.3   **Surface Versus Bulk Composition** (Ref. 5)

Bi:Mo atomic ratio versus bulk Bi:Mo ratio. (Points on this line with slope of 1 have the same bulk and surface composition.) However, as can be seen from the plot, the materials with bulk Bi:Mo ratios of between about 1.7 and 0.8 all have surfaces with a Bi:Mo ratio of about 1.

This means that, as Mo is added to Bi_2MoO_6, even in very small amounts, the Mo preferentially goes to the surface of the catalyst. Once the 1:1 Bi:Mo bulk ratio is reached (at $Bi_2Mo_2O_9$), addition of more Mo results in it preferentially going to the bulk to maintain a 1:1 surface ratio. The range of surface 1:1 ratio also corresponds to the range of active catalyst compositions, as measured by the reaction of butene to butadiene. The production of stable surface species across a range of bulk compositions occurs in many selective oxidation catalysts and accounts for their robust operation and long life even under harsh industrial conditions of high temperature and throughput.

3.2.4 *Promoters*

Another means of optimizing catalyst performance is by the use of promoters. Halogen, especially Cl, is a common promoter in many catalyst systems. In the AgO catalyzed epoxidation of ethylene to ethylene oxide, the addition of Cl (as was discussed in 3.1.8) improves selectivity to ethylene oxide (Fig. 3.2.4, left graph), which can be correlated with Cl coverage on the catalyst surface (Fig. 3.2.4, right graph).

3.2.5 *Oxidation Catalyst Synthesis*

We have seen that that unselective products are formed from bridging –O–Ag–O atoms (3.1.8). When Cl_2 is added to the feed, however, these unselective O atoms are replaced with Cl atoms, which do not react with ethylene. Thus selectivity is improved by replacement of unselective O–Ag–O sites with more stable Cl–Ag–Cl sites. The maximum selectivity achievable by the (unpromoted catalytic) mechanism is 5/6 or 83% (3.1.8). This is because the unselective O atoms must be removed by reaction with ethylene to form CO_2 before the active and selective sites can be regenerated. It is now possible to achieve selectivities in excess of the 83% limit for conventional AgO by the addition of Cl promoter to the catalyst.

Selective Oxidation $5\ H_2C=CH_2 + 5/2\ O_2 \longrightarrow 5\ H_2C\overset{\displaystyle{}}{\underset{O}{\diagup\!\!\diagdown}}CH_2$

Unselective Oxidation $H_2C=CH_2 + 3\ O_2 \longrightarrow 2CO_2 + 2H_2O$

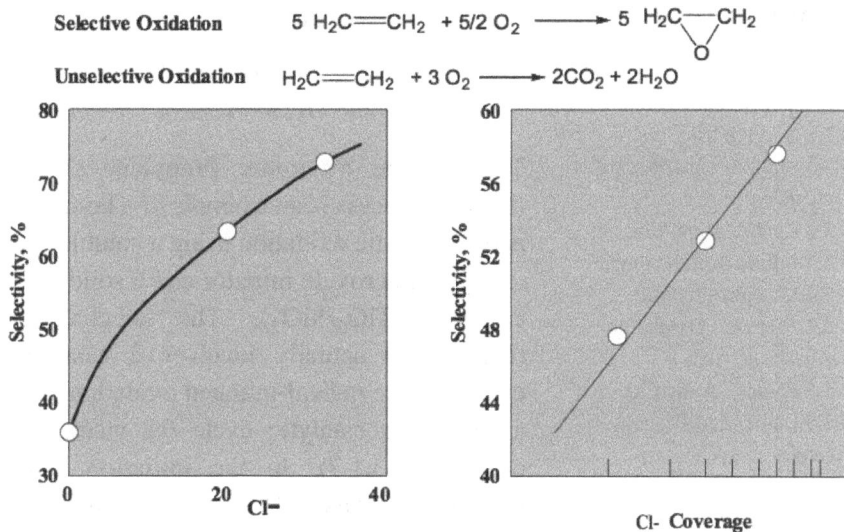

Fig. 3.2.4   Oxidation Catalyst Promoters (Ref. 1, reproduced with permission)

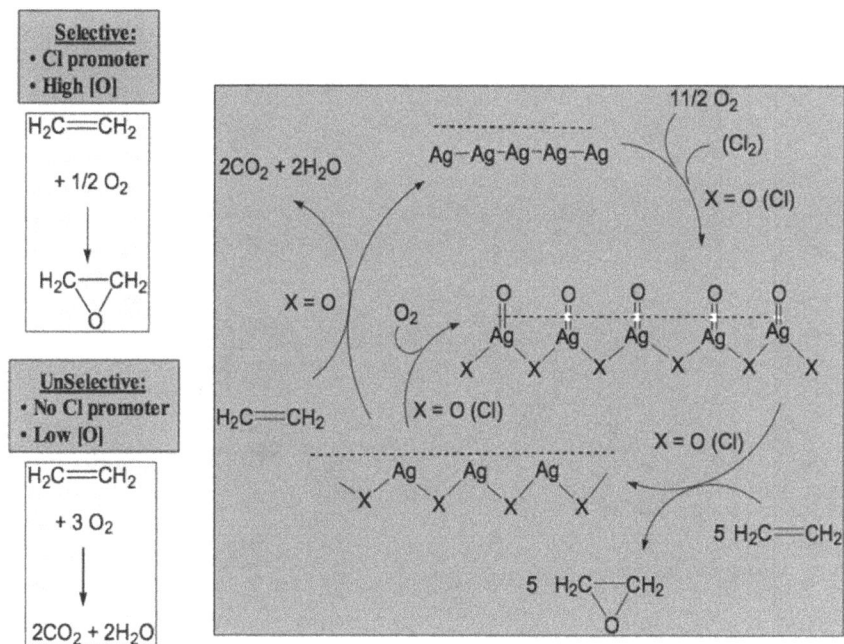

Fig. 3.2.5   Cl-Promotion (Ref. 1)

3.3 Major Industrial Processes

3.3.1 Epoxidation/SMPO
3.3.2 Acrylonitrile
3.3.3 Vinyl Acetate/Wacker Oxidation
3.3.4 Cyclohexanol/one
3.3.5 Adipic Acid/Nylon 6.6
3.3.6 Maleic Anhydride
3.3.7 Aromatic Oxidation — Terephthalic Acid
3.3.8 Phthalic Anhydride
3.3.9 Cumene to Phenol/Acetone

3.3 Major Industrial Processes

3.3.1 *Epoxidation — Styrene Monomer Propylene Oxide Process*

The Styrene Monomer Propylene Oxide (SMPO) process is an example of a homogeneous catalytic oxidation using a soluble Mo catalyst, a peroxide initiator and a solid acid catalyst (TiO_2/SiO_2). The mechanism (Fig. 3.3.1) actually involves 2 catalytic cycles: a free radical-initiated oxidation, and a molecular catalytic cycle (as were discussed in 3.1.7). In the initiation step, hydroxyl radicals (OH•), produced from the thermal peroxides decomposition, abstract a benzylic hydrogen from ethylbenzene to form the benzyl radical PhCH•CH$_3$. The benzylic

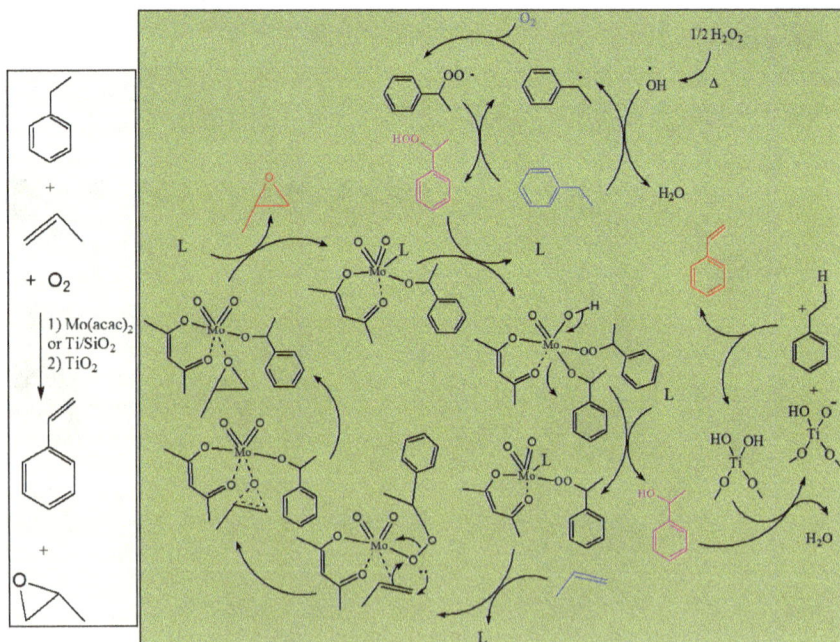

Fig. 3.3.1 SMPO Process (Ref. 6)

radical reacts with O_2 to form the benzylic peroxide radical, which abstracts a benzylic H from ethylbenzene to produce the benzylic hydroperoxide and reform the benzylic radical, thus propagating the radical chain. The hydroperoxide is the stoichiometric oxidant in the molecular catalytic cycle. In this molecular catalytic cycle, hydroperoxide reacts with a soluble Mo oxo species to form a Mo–peroxide complex, which oxidizes propylene to propylene oxide and a Mo benzylic alcohol complex. This complex is, in turn, oxidized by the benzylic hydroperoxide to reform the Mo peroxo species. In the process, the Mo benzylic alcohol complex undergoes ligand dissociation to produce the corresponding benzylic alcohol, which is dehydrated by a cationic mechanism catalyzed by the TiO_2 catalyst component to produce styrene.

Styrene is an important monomer used in polystyrene-based plastics and foams, ABS (acrylonitrile/butadiene/styrene) resins and SB (styrene/butadiene) rubber. Propylene oxide is a raw material for polyethers which are widely used as non-ionic surfactants and conditioners.

3.3.2 *Acrylonitrile*

Acrylonitrile is also a high-volume monomer used extensively in the plastics industry. It is produced commercially by a heterogeneous catalytic process known as ammoxidation (oxidation in the presence of ammonia) of propylene. One of these commercial catalysts is based on $Bi_2Mo_3O_{12}$ (α–bismuth molybdate). This mechanism (Fig. 3.3.2) is similar to the oxidation of propylene to acrolein except that N, instead of O, is inserted in the form of a Mo=NH species, which forms from NH_3 and a surface Mo=O group. Since this is a $6e^-$ oxidation, an additional H-abstraction is required compared to what occurs in the acrolein oxidation case (which is a 4e oxidation). As in the oxidation reaction to acrolein, the Mars–van Krevelen mechanism of bulk lattice participation and reoxidation are operative in catalytic ammoxidation of propylene over $Bi_2Mo_3O_{12}$.

3.3.3 *Vinyl Acetate/Wacker Oxidation*

The Wacker oxidation (Fig. 3.3.3) is an example of homogeneous molecular catalysis which uses Pd as the primary oxidation site and Cu

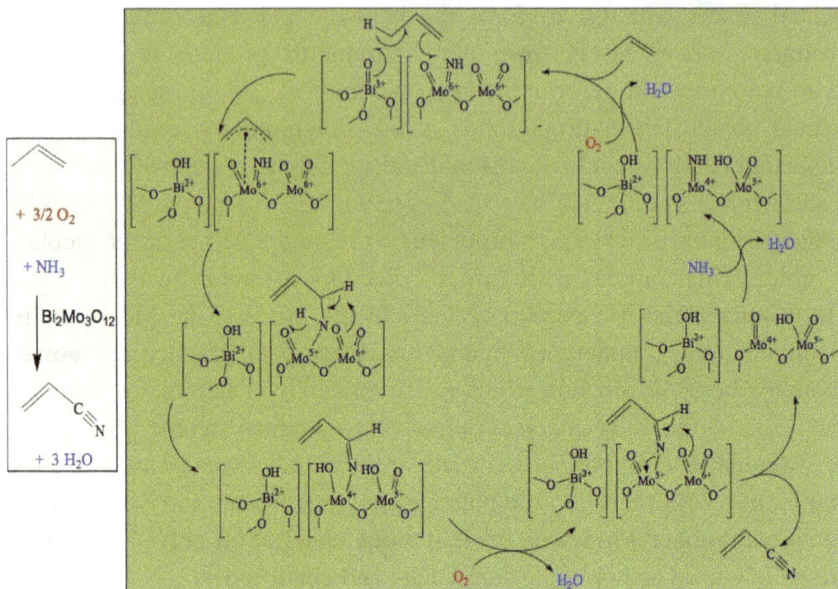

Fig. 3.3.2   Ammoxidation of Propylene to Acrylonitrile (Ref. 7, reproduced with permission)

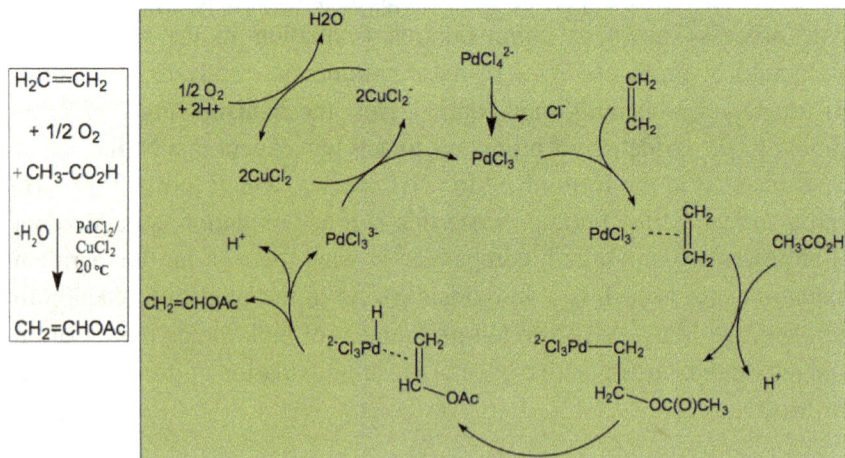

Fig. 3.3.3   Wacker Oxidation to Vinyl Acetate (Ref. 6, reproduced with permission)

as the reoxidation component, to produce vinyl esters or aldehydes from olefins. One of the most prominent reactions in this class is the production of vinyl acetate from ethylene, acetic acid and O_2. The Pd(2+) catalytic component, in the form of a tetrahedral (4-coordinate) $[PdCl_4]^{2-}$ salt, produces another tetrahedral complex by loss of Cl– and coordination of ethylene. Addition of acetic acid with 1e reduction of Pd, followed by –H elimination and another 1e reduction of Pd produces vinyl acetate and the Pd (0) species $[PdCl_2]^{2-}$. Cu (2+) as $CuCl_2$ reoxidizes the Pd(0) back to Pd (2+) (2 moles of Cu per mole of Pd) to start the cycle again, and the Cu(1+) formed in the process is oxidized back to Cu (2+) by molecular oxygen (O_2). In the process of forming vinyl acetate and Pd(0), 2 protons are produced, which are used in the Cu(1+) to Cu(2+) oxidation. Without the Cu, the reoxidation of Pd (0) is very slow and eventually the reaction stops as all the Pd (2+) is converted to Pd(0).

3.3.4 Cyclohexanol/one — Cyclohexane Oxidation

Cyclohexanone is the intermediate used in the production of cyclohexanone oxime, which in turn, is used to make nylon-6 (2.4.3). This is a free radical homogeneous mechanism (Fig. 3.3.4) catalyzed by Co, similar to what was discussed in 3.1.7. Hexane is activated by H-atom

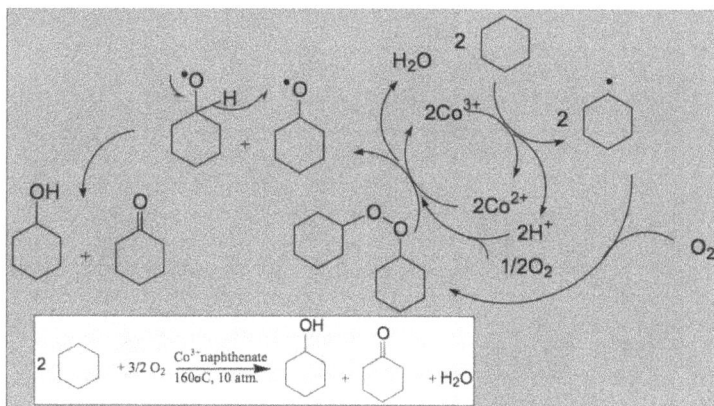

Fig. 3.3.4 Cyclohexane Oxidation (Ref. 6, reproduced with permission)

abstraction by Co (3+) to produce cyclohexyl radicals, which react with oxygen to form di-cyclohexyl peroxide. The weak O–O bond in this peroxide breaks thermally to form cyclohexyloxy radical, which disproportionates to a cyclohexane/cyclohexanone mixture (referred to as "ol/one"). Two protons (formed in the process of cyclohexane activation) and O_2 reoxidize Co (2+) to Co (3+) and H_2O to complete the catalytic cycle.

3.3.5   *Adipic Acid/Nylon 6,6*

The ol/one mixture is oxidized in a catalytic process shown in Fig. 3.3.5 to produce adipic acid (hexandioic acid), a key monomer used in production of nylon 6,6. This is a homogeneous catalytic process which uses nitric acid as the stoichiometric oxidant and a V(5+) or Cu (2+) catalyst. Adipic acid is converted to nylon 6,6 by condensation with

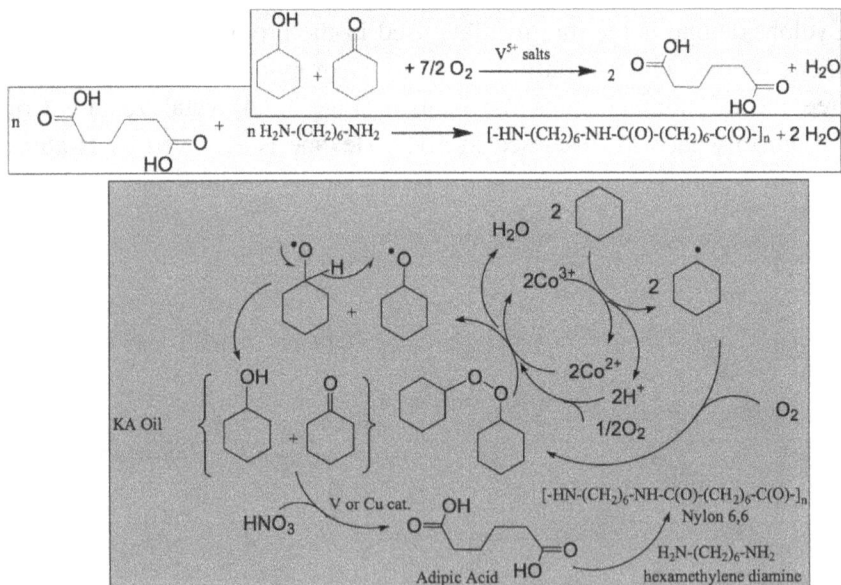

Fig. 3.3.5   Adipic Acid from Cyclohexane (Ref. 6, reproduced with permission)

hexamethylene diamine. Nylon 6,6 is an important polymer used in fibers and plastics.

3.3.6 *Maleic Anhydride*

Maleic anhydride is produced by a heterogeneous catalytic process from butane and air. This is one of the few examples of a selective paraffin oxidation process, which is used on a very large commercial scale.

3.3.6.1 Maleic anhydride mechanism

The mechanism has been well studied, but is still the topic of much discussion and speculation. A proposed mechanism is shown in Fig. 3.3.6.1, which attempts to fit the known features of this process, including those shown in Fig. 3.3.6.2.

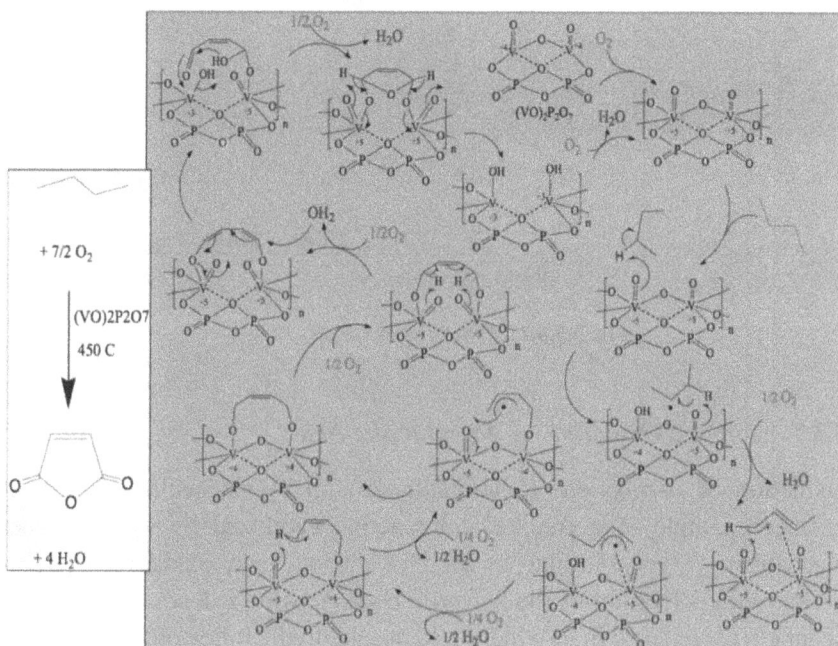

Fig. 3.3.6.1   Maleic Anhydride Mechanism (Refs. 22, 23)

3.3.6.2 O-utilization

These features include a $(VO)_2P_2O_7$ active phase, in which V(5+) and
V(4+) active species are involved in the catalytic cycle. While butadiene
is not a discrete desorbed intermediate, an adsorbed butadiene-like
intermediate may be formed. While the Mars–van Krevelen mechanism is
operative, experiments with $^{18}O_2$ indicate that only about 4 atomic layers
of lattice O are involved in the process, making it much more of a surface
reaction than for other heterogeneous oxidation catalysts, such as
$Bi_2Mo_3O_{12}$.

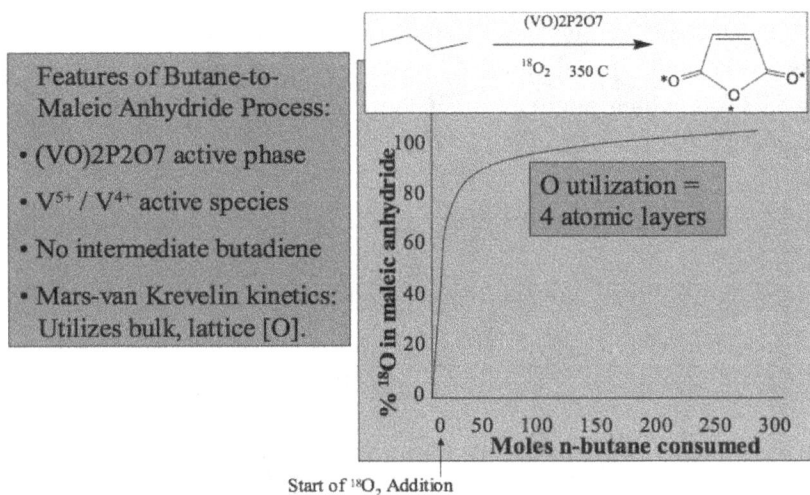

Fig. 3.3.6.2   **O-Utilization** (Refs. 22, 23)

3.3.7 *Aromatic Oxidation — Terephthalic Acid*

Oxidation of para-xylene to terephthalic acid by molecular oxygen is
another example of a free radical reaction catalyzed by Co(3+). The
mechanism is very similar to that for cyclohexane oxidation (3.3.4),
except with activation of the benzylic hydrogens of xylene. The process
occurs in two stages: first oxidation to the aldehyde, followed by aldehyde
oxidation to carboxylic acid as shown in Fig. 3.3.7. Terephthalic acid is an
important monomer used in production of plasticizers.

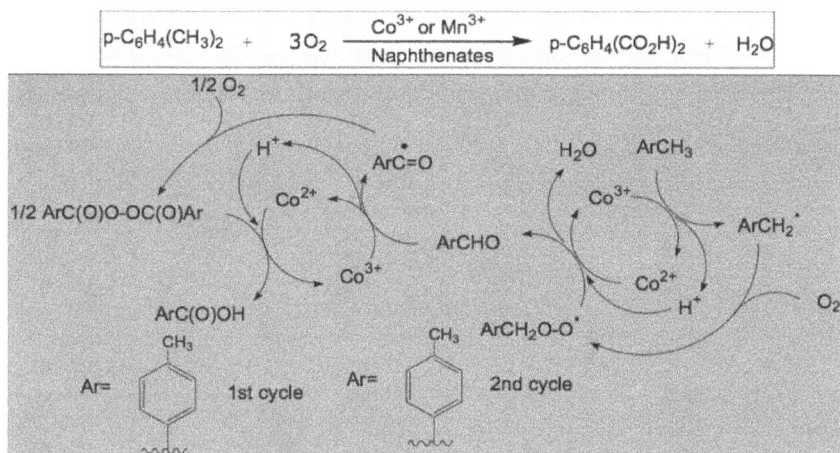

Fig. 3.3.7   Aromatic Oxidation — Terephthalic Acid (Refs. 5, 6)

3.3.8 *Phthalic Anhydride*

Phthalic anhydride production is analogous to maleic anhydride from butane. It uses the same catalyst and similar conditions and is thought to proceed by basically the same mechanism (Fig. 3.3.8) as the butane–tomaleic process.

3.3.9 *Aromatic Oxidation — Cumene to Phenol/Acetone*

Cumene to phenol is again a cocatalyzed free radical process similar to terephthalic acid (Fig. 3.3.9) and the "ol/one" (Fig. 3.3.4) process.

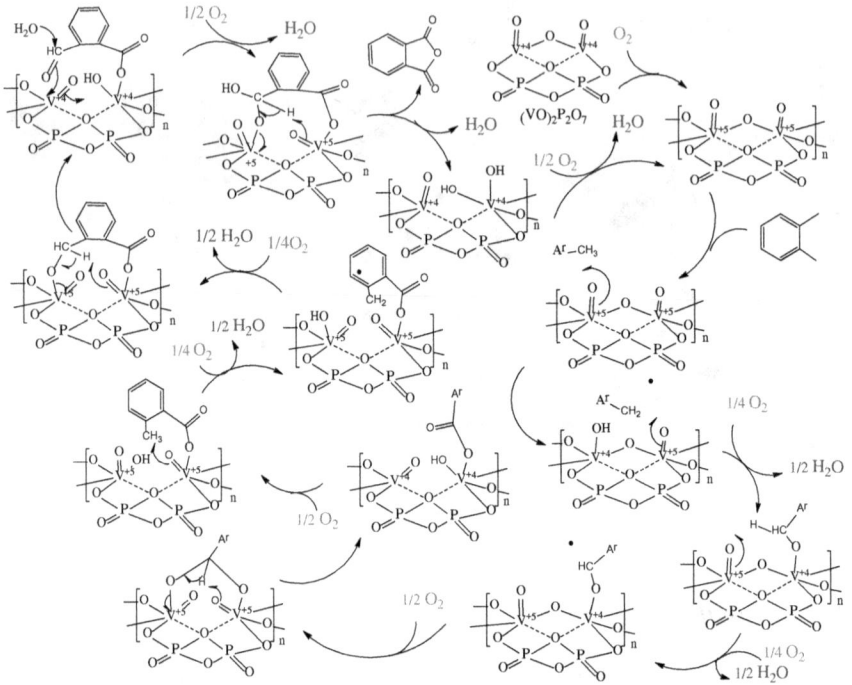

Fig. 3.3.8   Phthalic Anhydride (Ref. 18)

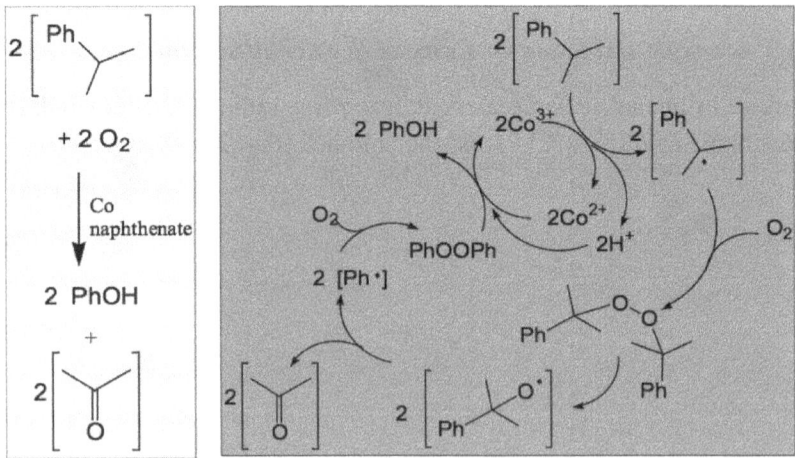

Fig. 3.3.9   Cumene to Phenol/Acetone (Ref. 18)

3.4 Problems

1. For the following molecules, state the oxidation level above the corresponding paraffin (P+__) in the blank space to the right of each structure:

(a)

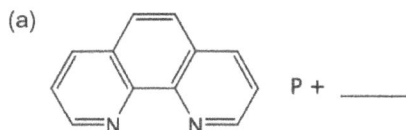

P + _____

(b)

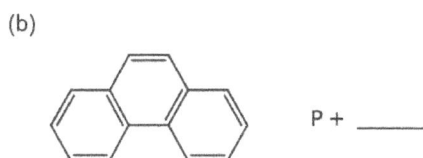

P + _____

(c)

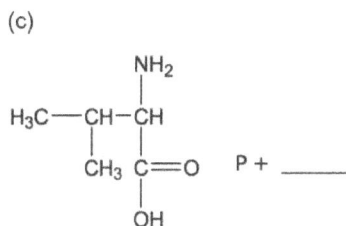

P + _____

(d)

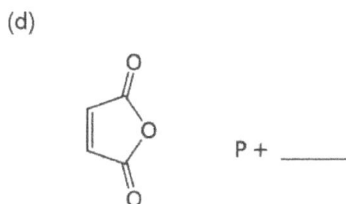

P + _____

(e)

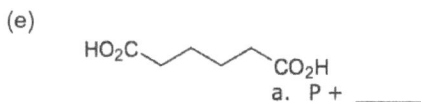

a. P + _____

(f)

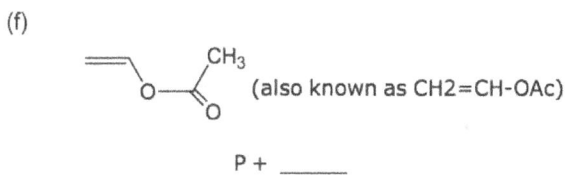

(also known as CH2=CH-OAc)

P + _____

2. For the following transformations, fill in the table below with the change (?) in the level of each molecule above the corresponding paraffin (Δ [P+ _]). Number of electrons in the oxidation (number of e–s), and the equivalent amount of O_2 that would be required (Number of O_2s):

a.

$CH_2=CH_2$ ⟶

b. $CH_2=CH-CH_2-CH_3 \rightarrow CH_2=CH-CH=CH_2$
c. $Ph-CH_3 \rightarrow PhCO_2H$
d. $CH_2=CH-CH_3 \rightarrow CH_2=CH-CN$

e.

$CH_2=CH-CH_3$ ⟶   —CH_3

f.

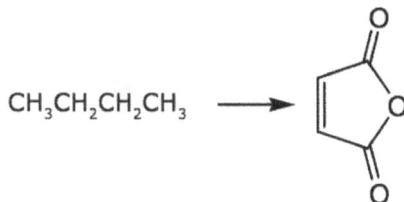

$CH_3CH_2CH_2CH_3$ ⟶

g. $CH_2=CH_2 \rightarrow CH_2=CH-OAc$

? [P + __]   No of e-'s   No. of O_2's

	? [P + __]	No of e-'s	No. of O_2's
a			
b			
c			
d			
e			
f			
g			

3. For each of the transformations in Problem 2, what are the actual oxidants, other reactants, catalysts that you would actually use to effect the reactions? Also indicate whether the catalysts are homogeneous or heterogeneous and select the mechanism from the list below the table. The reaction temperature you would use is given.

	Oxidant	Other Reactant	Catalyst	Reaction Temperature, °C	Mechanism*
a				320–380	
b				280–350	
c				110–120	
d				350–450	
e				90–20	
f				380–500	
g				90–110	

* HmM = homogeneous molecular; HmFR = Homogeneous Free Radical; HtS = Heterogeneous, Surface; HtB = Heterogeneous Bulk (Mars–van Krevelen).

4. Write a chemical equation for key activation steps in the mechanism for the transformations shown:

(a) Reaction in 2e using a soluble Mo catalyst and *tert*–Bu –OOH:

(b) Reaction in 2c using O_2 and Co naphthenate catalyst:

(c) Reaction in 2a using O_2 and a silver catalyst:

(d) Reaction in 2d using O_2 and a $Bi_2Mo_3O_{12}$ catalyst:

3.5 Answers to Problems

1.

(a)

P+ 12

(b)

P+ 10

(c)

P+ 4

(d)

P+ 7

(e)

P+ 6

(f)

P+ 2 (above the C2 paraffin — i.e., CH3CH3)

2.

	Δ [P+___]	Number of e−'s	Number of O2's
(a)	1	2	1/2
(b)	1	2	1/2
(c)	3	6	3/2
(d)	3	6	3/2
(e)	1	2	1/2
(f)	7	14	7/2
(g)	1	2	1/2

3.

	Oxidant	Other Reactant	Catalyst	Reaction Temperature, °C	Mechanism*
(a)	O_2	None	AgO	320–380	HtS
(b)	O_2	None	Bi_2MoO_6	350–450	HtB
(c)	O_2	None	Co–naphthenate	110–120	HmFR
(d)	O_2	NH3	$Bi_2Mo_3O_{12}$	350–450	HtB
(e)	H_2O_2 or ROOH	None	$Mo(acac)_2$	90–120	HmM
(f)	O_2	None	$(VO)_2P_2O_7$	380–500	HtB
(g)	O_2	HOAc	$PdCl_2/CuCl_2$	20–100	HmM

* HmM = homogeneous molecular; HmFR = Homogeneous Free Radical; HtS = Heterogeneous, Surface; HtB = Heterogeneous, Bulk (Mars–van Krevelen).

4.

(a) Reaction in 2e using a soluble Mo catalyst and *tert*-Bu –OOH:

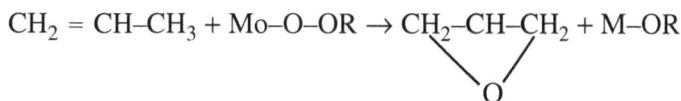

$$CH_2 = CH–CH_3 + Mo–O–OR \rightarrow CH_2–CH–CH_2 + M–OR$$

with O bridging.

(b) Reaction in 2c using O_2 and Co–naphthenate catalyst.:

$$PhCH_3 + Co (3+) \rightarrow PhCH_2\bullet + H+ + Co(2+)$$

(c) Reaction in 2a using O_2 and a silver catalyst:

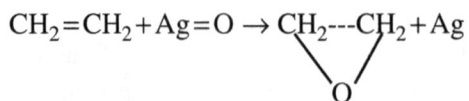

$$CH_2=CH_2+Ag=O \rightarrow \overset{CH_2\text{---}CH_2}{\underset{O}{\triangle}}+Ag$$

(d) Reaction in 2d using O_2 and a $Bi_2Mo_3O_{12}$ catalyst:

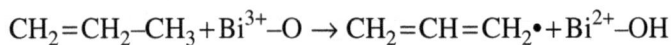

$$CH_2=CH_2\text{–}CH_3+Bi^{3+}\text{–}O \rightarrow CH_2=CH=CH_2\bullet+Bi^{2+}\text{–}OH$$

4

Polymerization Catalysis

Polymers account for the largest fraction of non-fuel chemical products produced by catalytic processes. Just the three highest volume polymers alone — polyethylene, polypropylene (PP) and polystyrene — account for about 35 billion pounds of product. Furthermore, many of the other high-volume chemicals produced from catalysis are monomers (many produced by acid or oxidation catalysts), which ultimately wind up in polymers by a catalytic polymerization reaction. So a large percentage of the materials produced by catalysis are, in one way or another, used for polymer production.

   The importance of polymerization catalysis is based on the breadth of products and applications that have been discovered and developed for polymers and plastics. It may be said that no other single class of materials has so influenced modern civilization. It is therefore not surprising that much of the industrial catalytic science and technology that has been developed in the past century has focused on polymerization catalysis. The structure of a polymer and the mechanism by which it is produced determines its properties and performance in a given application. The polymerization catalyst, in turn, is a key factor in determining polymer structure and properties. In this chapter is discussed the fundamentals of polymerization and the various mechanisms by which the breadth of polymers available today by modern industrial catalysts are produced.

As for other catalytic systems, control of product selectivity can be controlled by the selection of metal, ligands, solvent, single or multiple metal sites, and support particle. For cationic polymerization systems, the selection of counter anion is also important.

Among the major industrial polymers are polyethylene, and PP, olefin copolymers, oligomers, SBR, polyethoxylates and polyisobutylene. An interesting development in the production of branched polyethylenes from late transition metal systems is also discussed.

4.1 Concepts

The concepts of catalyst composition, thermodynamics, and mechanism are discussed in this section.

4.1.1 *Polymerization Catalyst Periodic Table*

Elements which dominate in polymerization catalysts (Table 4.1.1.1) include:

- Transition metals in their lower oxidation states,
- Elements which comprise strong Bronsted (halogen) and Lewis (B, Al) acids,

Table 4.1.1.1: **Polymerization Catalyst Periodic Table**

1a	2a	3b	4b	5b	6b	7b	8			1b	2b	3a	4a	5a	6a	7a	0
H																	He
Li	Be											B	C	N	O	F	Ne
Na	Mg											Al	Si	P	S	Cl	Ar
K	Ca	Sc	Ti	V	Cr	Mn	Fe	Co	Ni	Cu	Zn	Ga	Ge	As	Se	Br	Kr
Rb	Sr	Y	Zr	Nb	Mo	Te	Ru	Rh	Pd	Ag	Cd	In	Sn	Sb	Te	I	Xe
Cs	Ba	La*	Hf	Ta	W	Re	Os	Ir	Pt	Au	Hg	Tl	Pb	Bi	Po	At	Rn
Fr	Ra	Ac**															
*Lanthanides		Ce	Pr	Nd	Pm	Sm	Eu	Gd	Tb	Dy	Ho	Er	Tm	Y	Lu		
**Actinides		Th	Pa	U	Np	Pu	Am	Cm	Bk	Cf	Es	Fm	Md	No	Lw		

- Organoaluminum compounds and
- The above with P, S, N and O ligands.

The major polymerization catalyst types and the corresponding elements are given in Table 4.1.1.2. Each polymerization type is color-coded. The Ziegler–Natta/metallocene catalysts (blue) are the Group 4b, 5b, 6b and 7b transition metals, most notably Ti, Zr, Hf, V, Cr, Mo, W, Mn, Te and Re, in combination with an organo-aluminum compound.

Anionic catalysts (red) are hydrides of the alkali metals, typically Li, Na and K. Cationic catalysts (yellow) are acidic elements, especially B, Al, Ti, Si, Fe and Sn in high oxidation states. Free radical catalysts (purple) are commonly organic compounds with C–, N– or O–centered radicals. Ti and Al (in green) are unique in that they are common to both acidic (cationic) polymerization catalysts (in higher oxidation states) and to Ziegler–Natta/metallocene catalysts (in lower oxidation states). Hydrogen (orange) is present in both anionic catalysts (as H$^-$) and cationic catalysts (as H$^+$).

Table 4.1.1.2:   Polymerization Catalyst View of the Periodic Table

4.1.2. *Thermodynamics*

The thermodynamically favored products in polymerization reactions depend on whether H_2 is present (Fig. 4.1.2). In the absence of H_2, coke formation is favored, as it is for acid catalysts. If H_2 is present, hydrogenolysis, or the reductive cleavage of C–C bonds to eventually form methane, is most favored. These are the reactions which must be avoided to achieve high selectivity in polymerization reactions.

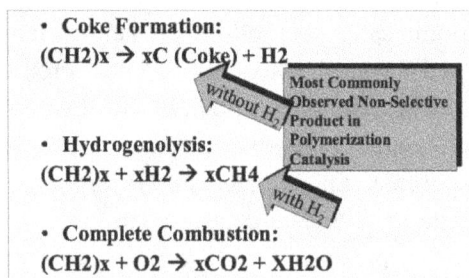

- **Coke Formation:**
  (CH2)x → xC (Coke) + H2

  without H₂

  Most Commonly Observed Non-Selective Product in Polymerization Catalysis

- **Hydrogenolysis:**
  (CH2)x + xH2 → xCH4

  with H₂

- **Complete Combustion:**
  (CH2)x + O2 → xCO2 + XH2O

Fig. 4.1.2   Thermodynamics

4.1.3 *Polymer Molecular Weight*

Polymerization reactions produce a distribution of molecular weight (Fig. 4.1.3.1).

Polymers are organic molecules which are composed of repeating molecular units or *monomers*. In general terms, the function of the polymerization catalyst is to lower the activation barrier for the formation of bonds between monomers. Thus, the catalyst serves as a means to activate the reaction of monomer with a growing polymer chain.

As the polymer grows, there is a finite probability that the growing chain will either react with more monomer to make a higher polymer, or undergo a chain transfer or termination reaction, which stops the chain from further growth and produces a polymer with specific number of monomeric units. The product of a polymerization reaction is, thus, not one discrete molecule with a precise molecular weight, but a distribution of molecules with an average molecular weight, which is determined by the probability of chain growth versus termination for any given intermediate chain length.

$$nM \longrightarrow (M)_n \xrightarrow{\;+\,M\;} (M)_{n+1} \xrightarrow{\;+\,M\;} (M)_{n+2} \xdashrightarrow{\;+\,mM\;} (M)_{n+m}$$

$$M_n \qquad\qquad M_{n+1} \qquad\qquad M_{n+2} \qquad\qquad M_{n+m}$$

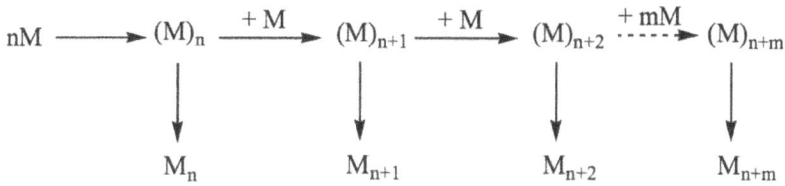

Fig. 4.1.3.1 Molecular Weight Distribution

The average molecular weight of polymer distribution can be defined in three ways, depending on which property of the molecular distribution is of interest (Fig. 4.1.3.2):

M_n = Number Average Molecular Weight
- $M_n = \Sigma n_i * M_{wi} / \Sigma n_i$
- Used to represent chemical stoichiometry

M_w = Weight Average Molecular Weight
- $M_w = \Sigma n_i * (Mw_i)^2 / \Sigma n_i * Mw_i$
- Used to represent physical properties

M_z = z-average Molecular Weight
- $M_z = \Sigma n_i * (Mw_i)^3 / \Sigma n_i * (Mw_i)^2$
- Used to represent mechanical properties

M_w/M_n = Polydispersity or Molecular Weight

Fig. 4.1.3.2 Molecular Weight Distribution

Distribution. This is a measure of the breadth of the polymer distribution. A polydispersity of 1 means that every molecule has the same molecular weight.

1. *Number Average Molecular Weight (M_n).*
This is defined as the summation of the molecular weights of each polymeric molecule multiplied by the number of that molecule, and divided by the total number of molecules. This is the average stoichiometric molecular weight, which is the average number of grams in a mole of the material.

2.   *Weight Average Molecular Weight* (M_w).

This is defined as the summation of the square of the molecular weights of each polymeric molecule multiplied by the number of that molecule, and divided by the summation of the molecular weights of each polymeric molecule multiplied by the number of that molecule. The physical properties of the polymer, such as the temperature at which the polymer undergoes phase transitions (for example, glass transition temperature, T_g) are determined by M_w.

3.   *Z-Average Molecular Weight* (M_z).

This is defined as the summation of the cube of the molecular weights of each polymeric molecule multiplied by the number of that molecule, and divided by the summation of the square of the molecular weights of each polymeric molecule multiplied by the number of that molecule. The mechanical properties of the polymer, such as strength (e.g., Young's modulus) and toughness are determined by M_z.

Consider a polymer of formula $(CH_2)_x$ with the distribution of polymer molecules shown in Table 4.1.3.3. The distribution is fairly narrow, with most of the molecules being the pentamer (containing 5 monomeric units) with a molecular weight of 70. Thus, one would expect the average number molecular weight (M_n) to be about 70, with a polydispersity close to 1.0. The actual figures are 65.8 and 1.08, respectively.

Now consider the same polymer but with half of the number of moles of pentamer (Table 4.1.3.4). The distribution is now broader, with the distribution of molecules skewed toward lower molecular weights. Thus,

For a polymer of the formula: $(CH_2)_x$

Table 4.1.3.3:   Molecular Weight Example 1

X	Molecular Weight	n (No. of moles)	n*MW	n*(MW)²
1	14	1	14	196
2	28	2	56	1568
3	42	3	126	5292
4	56	5	280	15680
5	70	20	1400	98000
6	84	3	252	21168
7	98	2	196	19208
8	112	1	112	12544
Σn = 37		Σn*MW= 2436	Σn*(MW)²= 173656	
Mn = 2436/37 = 65.8	Mw = 173656/2436 = 71.3	Mw/Mn = 1.08		

For a polymer of the formula: $(CH_2)_x$

Table 4.1.3.4:   Molecular Weight Example 2

X	Molecular Weight	n (No. of moles)	n*MW	n*(MW)²
1	14	1	14	196
2	28	2	56	1568
3	42	3	126	5292
4	56	5	280	15680
5	70	10	700	49000
6	84	3	252	21168
7	98	2	196	19208
8	112	1	112	12544
$\Sigma n = 27$		$\Sigma n*M_w = 1736$	$\Sigma n*(M_w)^2 = 124656$	
$M_n = 1736/27 = 64.3$	$M_w = 124656/1736 = 71.8$		$M_w/M_n = 1.12$	

one would expect the average number molecular weight (M_n) to be < 67.7, with a polydispersity > 1.05. The actual figures are 54.3 and 1.17, respectively. Thus, the number average molecular weight (M_n) and the polydispersity (M_w/M_n) together provide a good chemical description of the polymer: its average chemical molecular weight and the breadth of the distribution of molecules in the polymer.

4.1.4 *Polymerization Reaction Mechanisms*

The high volume monomers and the corresponding reaction types:

- Anionic
- Cationic
- Free Radical
- Metal Oxide/Coordination

that are used commercially to produce the corresponding polymers are summarized in Table 4.1.4. These will be discussed in detail in the rest of this chapter.

4.1.5 *Polymerization Mechanisms: Elementary Steps*

The major steps which are common to all four types of polymerization reactions are illustrated in Fig. 4.1.5 by the cationic polymerization of styrene. These are as follows:

Initiation: The catalyst initiates the reaction by production of a reactive monomer intermediate. In this case (cationic polymerization), the reactive

Table 4.1.4:   Polymerization Reaction Mechanisms (Ref. 24, reproduced with permission)

Monomer		Major Application	Anionic	Cationic	Free Radical	Metal Oxide / Coordination
Ethylene	CH2=CH2	Plastics			X	X
Propylene	CH3CH=CH2	Plastics			X	X
Isobutylene	CH2=C(CH3)2	Rubber Goods		X		
Butadiene	CH2=CH-CH=CH2	Rubber Goods	X		X	X
Isoprene	CH2=C(CH3)CH=CH2	Rubber Goods	X		X	X
Styrene	CH2=CHPh	Plastics	X	X	X	X
Nitroethylene	CH2=CHNO2	Intermediate	X			
Vinyl Ethers	CH2=CHOR	Intermediate		X		X
Vinyl Pyrrolidone	(structure)	Performance Additives		X	X	
Methyl Methacrylate	CH2=C(CH3)CO2CH3	Coatings	X		X	X
Methyl α-cynanoactrylate	CH2=C(CN)CO2CH3	Adhesives	X		X	
Acrylonitrile	CH2=CHCN	Plastics	X		X	

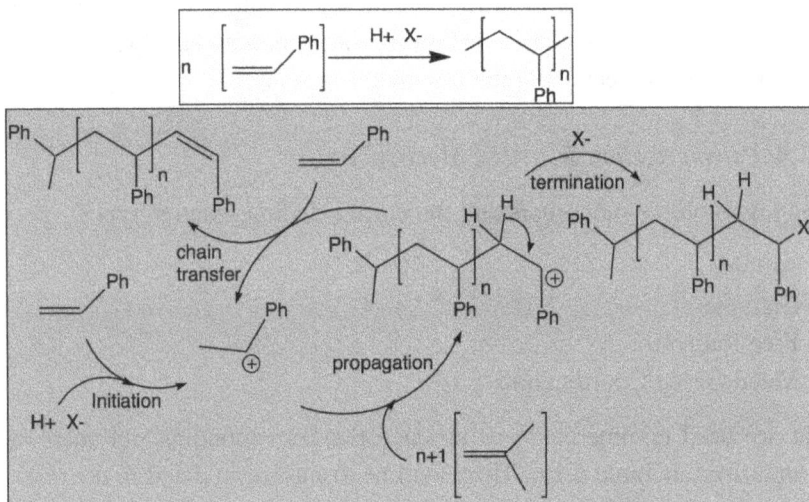

Fig. 4.1.5   Polymerization Mechanisms: Elementary Steps

intermediate is a carbenium ion, but it could also be a free radical, an anion, or a organometallic coordination complex.

Propagation: Monomer reacts with the reactive intermediate to produce a growing polymer chain with a reactive end group.

Chain transfer: Hydrogen, as H^+, H^- or $H\bullet$, is transferred between the reactive end group and a monomer to regenerate the reactive monomer intermediate. This process continues the reactive catalytic chain and, along with the propagation step, is the second step in the catalytic cycle. One initiation event generates many propagation steps, generating many polymer molecules, and the chain transfer steps generates the reactive monomeric intermediate.

Termination: Hydrogen, as H^+, H^- or $H\bullet$, or some other reactive species (X^+, X^-, X^0) is transferred between the reactive intermediate and a species other than the monomer. This process also produces polymer, but terminates the reactive chain. In this process, the "catalyst" that is added to the reaction (in this case HX) is actually an "initiator", because it initiates the production of the reactive monomer intermediate. One initiation event generates many propagation steps and generates many polymer molecules. However, after termination, another initiator molecule is required to produce more polymer.

4.1.6 *Anionic Polymerization*

The rest of this section is devoted to discussion of the catalysts and monomers used in each of the four major reaction types. First discussed is anionic polymerization. Anionic catalysts include Na and K metal and metal hydrides, sodium methoxide, potassium diphenyl azide, and Li, K and Na alkyls. Electron-poor monomers (such as acrylonitrile) and others which form stable anionic intermediates (including butadiene) are polymerized by anionic catalysts. Polymers of butadiene are known as *synthetic rubber*. The major commercial product is a polybutadiene rubber known as Krayton® rubber, produced by Shell (Fig. 4.1.6.1).

The basic mechanism for anionic polymerization is illustrated in Fig. 4.1.6.2 for metal-catalyzed polymerization of butadiene:

- *Initiation* by electron transfer from the catalyst (metal atom) to the monomer (butadiene) to form the reactive monomer intermediate (radical anion), which dimerizes to form a C_8 intermediate with a reactive anionic center on both ends.

Catalysts	Monomers
Na or K (in toluene, naphalene or liquid NH3	styrene, butadiene, stilbene
KNH2, NaNH2	styrene, butadiene
NaOCH3	styrene, acrylic esters, acrylonitrile, vinyl ketones, epoxides
Ar2N⁻K⁺	2-cyano-1,3-butadiene
RLi, RK, RNa	acrylonitrile, styrene, dienes

**Major Commercial Product:
Krayton Polybutadiene Rubbers
(Shell)**

Fig. 4.1.6.1   Anionic Polymerization (Refs. 24–29)

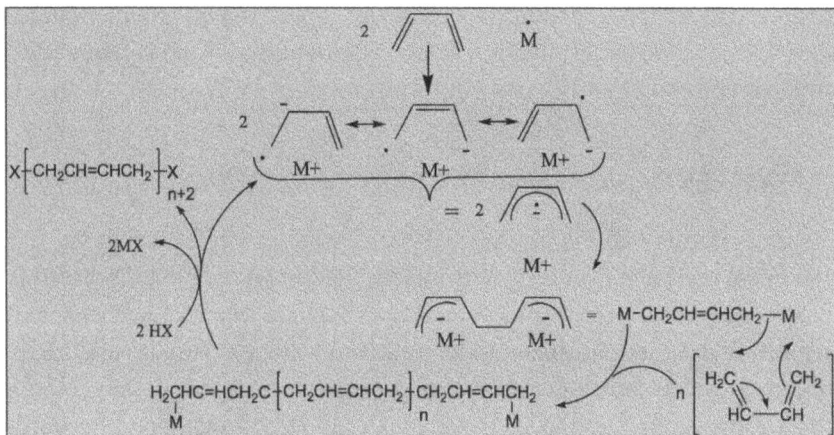

Fig. 4.1.6.2   Anionic Polymerization (Refs. 24, 25)

- *Propagation* by reaction of the anionic centers with monomer to form a polymer with anionic reactive end groups.

After all, monomer is converted to polybutadiene with metal anionic end groups, HX (e.g., HCl) is added to produce the polybutadiene product (with X end groups). In this case, since there is no chain transfer, each metal atom produces one polymer molecule. The molar ratio of metal to butadiene required is therefore 1: (M_n of the polybutadiene), which is typically in the 100,000's.

4.1.7 Cationic Polymerization

Next discussed is cationic polymerization (Fig. 4.1.7). Cationic catalysts include Lewis acids (BF_3, $AlCl_3$), Bronsted acids (H_2SO_4, HF) and salts of non-coordinating anions (Ph_3C+BF_4-). Monomers, such as isobutylene and styrene, which form stable carbenium intermediates are polymerized by cationic catalysts. Polyisobutylene is one of the largest volume monomers that is produced by cationic polymerization catalysts.

The mechanism of cationic polymerization of isobutylene shown in Fig. 4.1.7 has the same basic steps as the cationic polymerization of styrene as shown in Fig. 4.1.5: initiation by H^+ transfer, in this case, to form t-butyl cation, propagation to form the growing polymer chain, and chain transfer of H^+ from the growing polymer chain to isobutylene to form polyisobutylene with an olefin end group and regenerate the active tert-butyl cation.

Table 4.1.7: Cationic Polymerization Catalysis (Ref. 24)

	Catalysts	Monomers
Lewis Acids	BF3 (w/H2O, ROH)	2-butene, isobutylene
	AlCl3, AlBr3 (w/H2O, ROH)	isobutylene, styrene
	SnCl4	styrene
	TiCl4	propylene
	FeCl3 (with HCl)	styrene
Bronsted Acids	H2SO4, KHSO4	styrene, propylene (low MW)
	HF	styrene
	HClO4	styrene
	Cl3CO2H	α-methylstyrene
	(NH4)2.5H0.5PW12O40	isobutylene
Salts of Non-coordinating Anions	Ph3C+BF4-	styrene
	Ph3C+ (C6F5)4B-	isobutylene
	Ph3C+SbCl6	styrene
	(C2H5)3O+BF4-	styrene, propylene

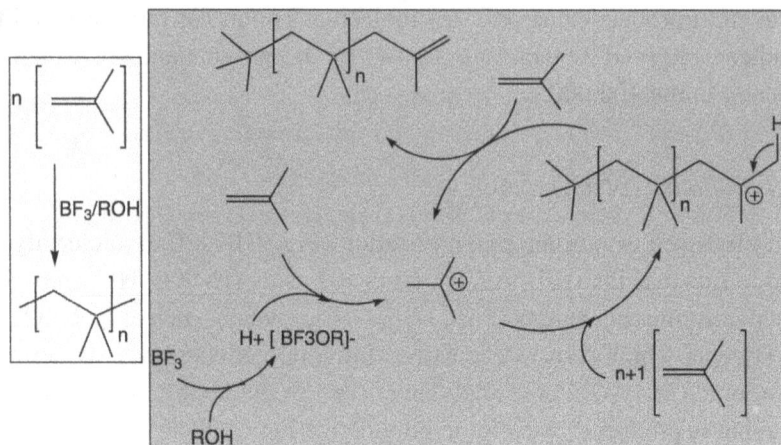

Fig. 4.1.7   Cationic Polymerization: Polyisobutylene

4.1.8 *Metal Oxide/Coordination Catalysis*

Next discussed is metal oxide and coordination-type catalysts (Fig. 4.1.8). These catalysts include *Ziegler–Natta* catalysts, *metallocene* catalysts and *metal oxide* catalysts. Ziegler–Natta catalysts are produced by the combination of a Lewis acid and an aluminum alkyl, and are used to produce high volume polyolefins such as polyethylene. *Metallocenes*, which are cyclic organometallic complexes of a rare earth element, usually Zr, and are also used to produce polyolefins, as are metal oxide catalysts.

Ziegler–Natta, metal oxide and coordination catalysts are formed from three basic components:

- A catalytic metal center
- A hydrogen or alkyl donor
- A catalyst support

Prominent examples of these components for Ziegler–Natta, metallocene and the chromium oxide "Phillips" catalyst are shown in Table 4.1.8.

Olefin Polymerization Catalysts	Catalysts	MW Distribution	Active Life	Incorporation of comonomers	End group
Ziegler–Natta–type	MClx + Al(CH2R)xCl3−x	narrow	intermediate	low	saturated
Examples	TiCl4/Al(C2H5)2Cl/MgCl2 VCl3/THF/AlEt3Cl/SiO2				
Metallocene	L1L2MXY M=Zr, Ti, Hf; L1=Cp, L2=organic ligand; X, Y=halogen	very narrow	long	high	unsaturated
Examples	Cp2ZrCl2/(CH3–Al–O)n Cp2Cr				
Metal Oxides	MOx/support	intermediate-broad	very long	high	unsaturated
Examples	CrO3/SiO2+H2 (Phillips) VO(OR)3/R2AlX				

Fig. 4.1.8 **Metal Oxide/Coordination Catalysis** (Refs. 24, 30)

Table 4.1.8: **Metal Oxide/Catalyst Formulation**

Catalyst Formula	Catalytic Metal Center	H- or R-Donor Activator	Catalyst Support
Ziegler–Natta	$TiCl_4$, VCl_3	$Al-Et_2Cl$	$MgCl_2$ or SiO_2
Metallocene	$ZrCp_2Cl_2$	$(Me-Al-O-)n$	None
Phillips	CrO_3	H_2	SiO_2

4.1.9 *Coordination Catalysis: Ziegler–Natta*

A representative Ziegler–Natta mechanism is shown in Fig. 4.1.9. Ziegler–Natta catalysts are formed from (i) a transition metal compound MX_n, where M-Ti or V, X is halogen and $n = 3$ or 4 and (ii) an organome-tallic compound such as AlR_3 or AlR_2X. One of the most commonly used systems is based on $TiCl_3$ and $AlEi_2Cl$.

The activation of catalyst occurs by donation of an alkyl from the aluminum alkyl to the metal center (M). The resulting metal alkyl forms an olefin complex and, subsequently, the corresponding metal alkyl by olefin insertion. Additional olefin reacts with this metal alkyl to grow the alkyl chain. This polymeric metal alkyl can undergo chain termination by reaction with hydrogen or more aluminum alkyl to form polyethylene

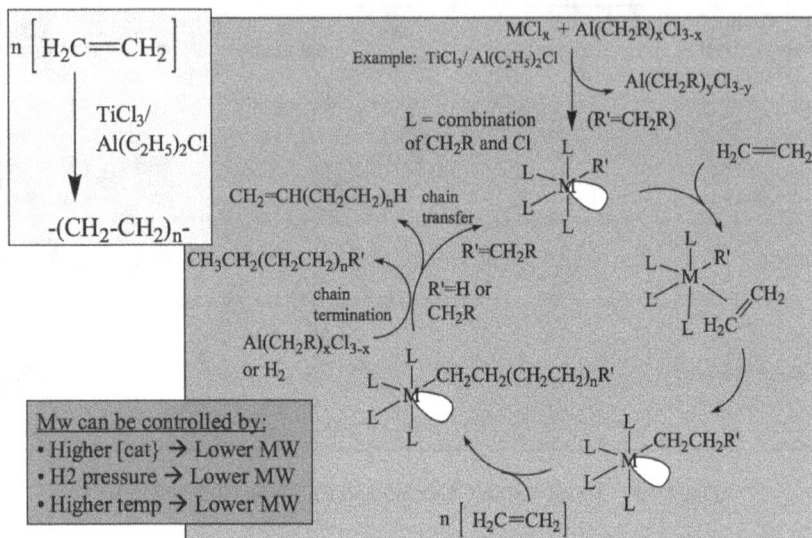

Fig 4.1.9   Coordination Catalysis: Ziegler–Natta (Refs. 5, 25–29, 31)

with a saturated hydrocarbon end group. Alternatively, it can undergo chain transfer to reform the metal alkyl and polyethylene with an unsaturated end group.

The molecular weight can be controlled by increasing concentration of catalyst or hydrogen, both of which decrease chain growth by increasing termination reactions. Higher temperature also decreases molecular weight by favoring chain transfer over chain growth.

4.1.10 Coordination Catalysis: Metallocene

In metallocene catalysis (Fig. 4.1.10), the active species is a Zr^+ complex, which is formed by donation of a CH_{3-} from methylaluminoxane (MAO) to $ZrCl_2$ with loss of 2 Cl^-. The Zr^+ forms an olefin complex, and then the corresponding metal alkyl; repetition of this process forms the metal alkyl, where the alkyl is the growing polymer chain. This can undergo β-H elimination to form polymer with an olefin end group and a metal hydride, which can activate more olefin. This constitutes a chain transfer step. Alternatively, the polymeric complex can react with MAO to produce a

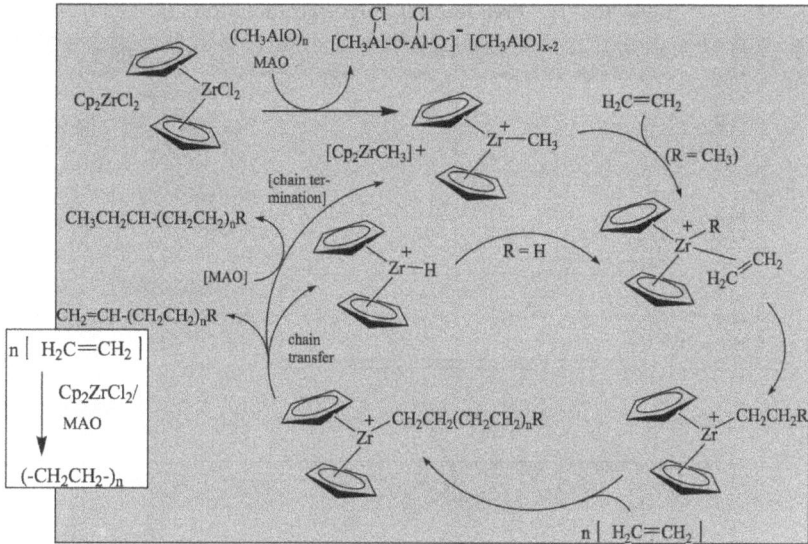

Fig. 4.1.10   Coordination Catalysis: Metallocene (Refs. 1, 30)

saturated olefin. The latter terminates the chain, but reforms the active $Cp_2Zr(CH_3)^+$ species. In both cases, only one mole of Zr is needed to produce many moles of polymer molecules. For chain transfer, only one mole of MAO produces many moles of polymer, but for chain termination, one mole of MAO is needed for every mole of polymer. Again, for termination mechanisms, on average, the ratio of catalyst (initiator) to monomer is $1:M_n$ of the polymer, which for metallocenes can be on the order of 10^5 or 10^6.

4.1.11 *Free Radical Polymerization*

Next discussed is free radical polymerization catalysis/initiators (Table 4.1.11). These catalysts include carbon-centered radicals, like those formed by thermolytic loss of N_2 from azo compounds like AIBN, and from O-centered radicals, like those formed from thermal cleavage of the O–O bond in DPB, or Perca dox-16. Monomers include those which form stable free radicals, such as styrene, butadiene acrylonitrile and maleic anhydride.

Table 4.1.11:   Free Radical Polymerization (Ref. 24)

	Catalysts	Monomers
AIBN		styrene, maleic anhydride, acrylonitrile, butadiene co-polymers
DPB		styrene, maleic anhydride, acrylonitrile, butadiene co-polymers
Percadox-16		styrene, maleic anhydride, acrylonitrile, butadiene co-polymers

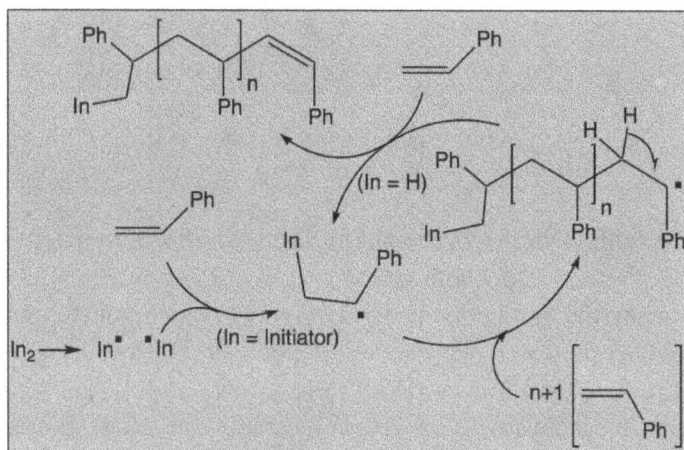

Fig. 4.1.11   Free Radical Styrene

Free radical polymerization, like cationic polymerization, occurs by the three basic steps: initiation, propagation and chain transfer and/or termination. As is illustrated in Fig. 4.1.11 for the free radical polymerization of styrene, initiation occurs by reaction of monomer with initiator to form a monomer radical species, which adds more monomer (propagation) to form a polymeric radical. This, in turn, can undergo H• transfer from radical polymer to monomer to form an activated radical monomer and the polymer (chain transfer process).

4.1.12 *Living Polymerization*

In a process known as *living free radical polymerization*, chain transfer and chain termination do not occur. Instead, if the initiator is a stable radical, like a nitroxide ($R_2NO\bullet$), it reversibly forms a stable product with the growing radical polymer. Thus, all available monomers react until it is consumed to form the initiator–polymer complex. The process is known as living polymerization, since the initiator–polymer species remains active (or living) and can continue to react with monomer as long as monomer is present. In the case shown in Fig. 4.1.12.1, the product has an initiator group on both ends of the polymer. Polymers which have chemical functionality on both ends of the molecule are known as *telechelic*. Thus, living free radical polymerizations produce telechelic polymers. It also produces molecules of molecular weight proportional to the concentration of monomer. If more monomer is added, the molecular weight continues to increase because the ends of the resulting polymer are *living*.

This effect is illustrated in Fig. 4.1.12.2 in a plot of M_n versus % monomer conversion. In typical chain-growth polymerization reactions, like the ones we have seen previously in this chapter, the average molecular

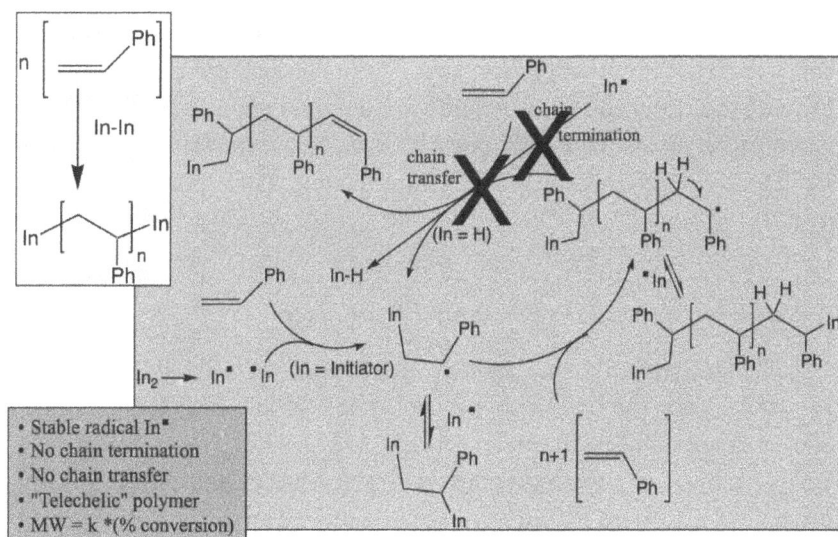

Fig. 4.1.12.1 Living Polymerization

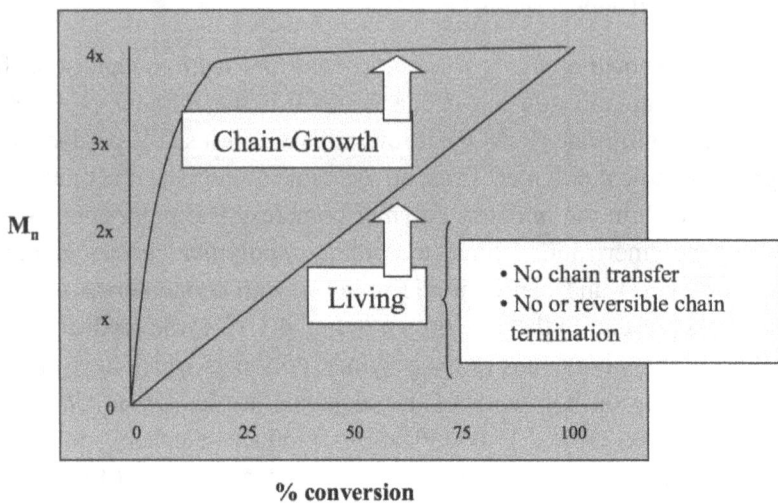

% conversion

Fig. 4.1.12.2 Living Polymerization (Refs. 25–29)

weight formed reaches a constant as the polymerization reaction reaches a steady state of propagation to chain transfer/termination reactions. In living polymerization, however, the molecular weight continues to rise throughout the course of the polymerization reaction, since there are no chain transfer or termination reactions. The stable initiator remains associated with the reactive end of the growing polymer chain, but available for reaction with all available monomer by dissociation of stable free radical (for example, $R_2NO\bullet$, as in 4.1.11) from the end of the growing polymer chain.

4.1.13 *Living Polymerization: TEMPO Initiator*

The mechanism of free radical polymerization is illustrated in Fig. 4.1.13.1 for the tetramethyl pyrrolidine N-oxide (TEMPO)-initiated copolymerization of methyl methacrylate and styrene. The stable N-oxide TEMPO radical remains associated with the reactive end of the growing polymer chain. Thus, living polymerization can be used to make copolymers with specific lengths of various polymer segments, in this case methacrylate and styrene, within the overall polymer.

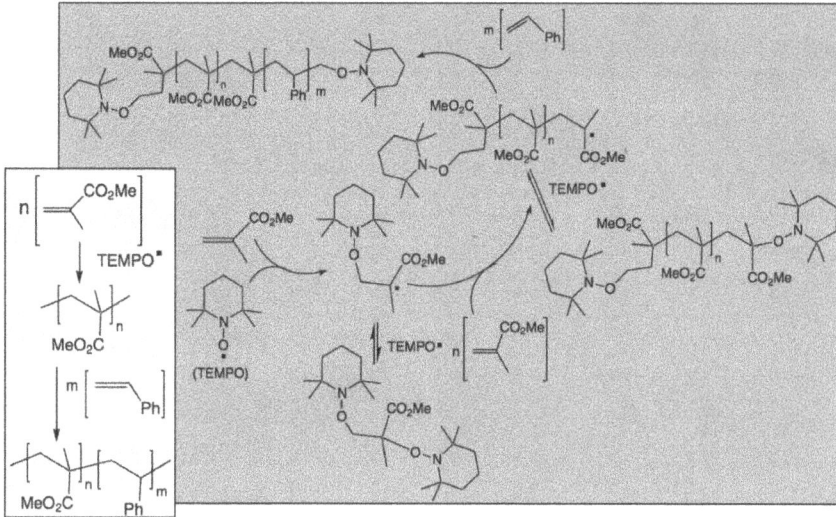

Fig. 4.1.13.1 Living Copolymerization

In this mechanism, chain transfer and termination are prevented, but the end of the growing polymer chain remains available for reaction with monomer by reversible dissociation of the TEMPO–polymer bond.

Developments in radical living polymerization include Atom Transfer Radical Polymerization (ATRP) (Ref. 32) and Reversible Addition-Fragmentation Chain Transfer (RAFT) (Ref. 33), which can be used to produce block copolymers with specific monomer compositions with a range of functionalities and architectures.

Radical initiated copolymerization of styrene and methyl methacrylate in two stages using the RAFT chain transfer agent [S = C(X)SR] to produce a methacrylate/styrene block copolymer is illustrated in Fig. 4.1.13.2. As for the TEMPO case, the length of the two polymer blocks is controlled by the amount of monomer that is added to each stage.

4.2 Polymerization Catalyst Design

Next discussed are some of the ways in which polymerization catalysts are designed for a specific process, as covered in this section. As for the other catalyst systems, metals substitution, choice of ligand, use of

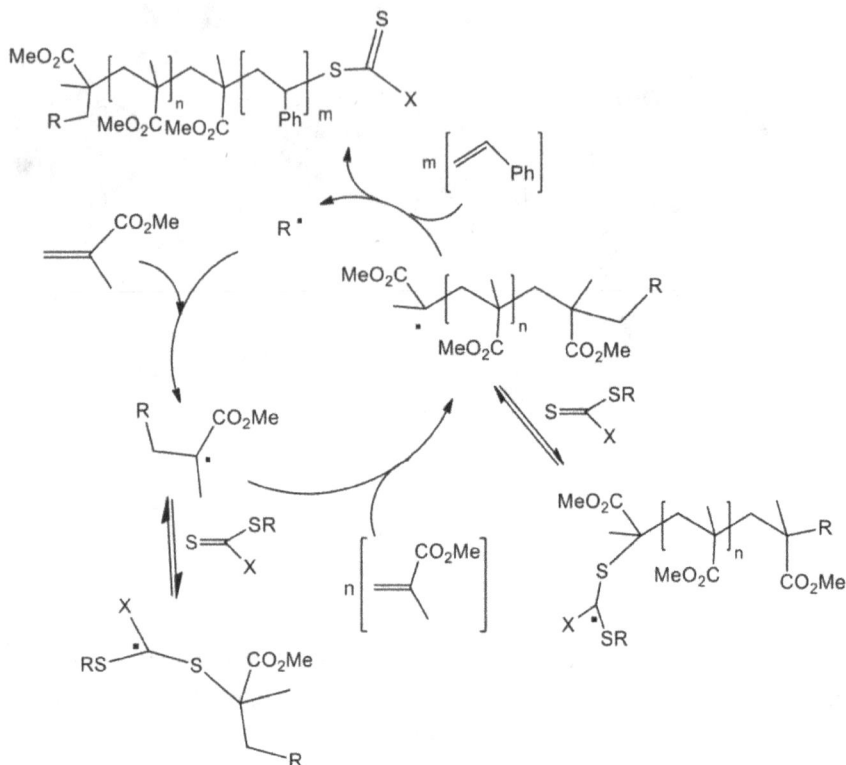

Fig. 4.1.13.2   RAFT Polymerization of Methyl Methacrylate and Styrene (Ref. 33)

non-coordinating anions, and reaction at single versus multiple catalytic sites are important ways to control catalytic performance.

4.2.1 *Metals Substitution: PP Isomers*

The selection of the catalyst metal center as a means to control selectivity is illustrated by the polymerization of propylene (Fig. 4.2.10). Three stereo-isomers are possible: atactic (random distribution of stereocenters), syndiotactic (alternating stereo centers) and isotactic (alike stereo centers). In general, the isotactic isomer has the most desirable physical properties, such as tensile strength, thermal and scratch resistance and optical properties. Early PP catalyst such as $TiCl_3/Et_2AlCl$ as tensile strength, thermal and scratch resistance and optical properties.

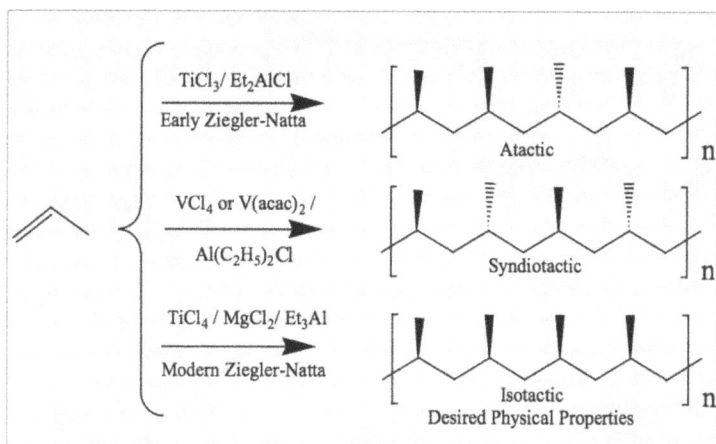

Fig. 4.2.1 Metals Substitution: PP Isomers (Ref. 1)

These catalysts contain $AlCl_3$ impurities and several types of catalytic sites, which undergo monomer insertion to produce a range of different stereochemistries, giving rise to the atactic backbone. V-based catalysts such as VCl_4 or $V(acac)_2/AlEt_3$ produce the syndiotactic isomer (Fig. 4.2.1). Modern Ziegler–Natta catalysts use very small crystallites of $MgCl_2$ as support for $TiCl_4$, which is free of $AlCl_3$ impurities and is reduced by $AlEt_3$ to give a higher uniformity of isotactic-producing catalyst sites. The differences in stereoselectivity between the V and modern Ti systems can be understood based on the orientation of the monomer insertion steps and the most favorable orientation of methyl and organometallic groups in the V-centered versus the Ti-centered intermediates.

4.2.2 *V-Based PP Catalysts: Syndiotactic*

For the V-based intermediate, the polymerization proceeds by 2,1 migratory insertion (Eq. 4.2.2):

$$V–R + CH_2 = CHCH_3 \rightarrow V–CH(CH_3)–CH_2–R. \qquad (4.2.2)$$

In other words, the methyl group in the vanadium alkyl complexes is on the carbon directly attached to the metal. The 2,1 addition in V-based PP catalysts has been demonstrated by termination experiments, which show

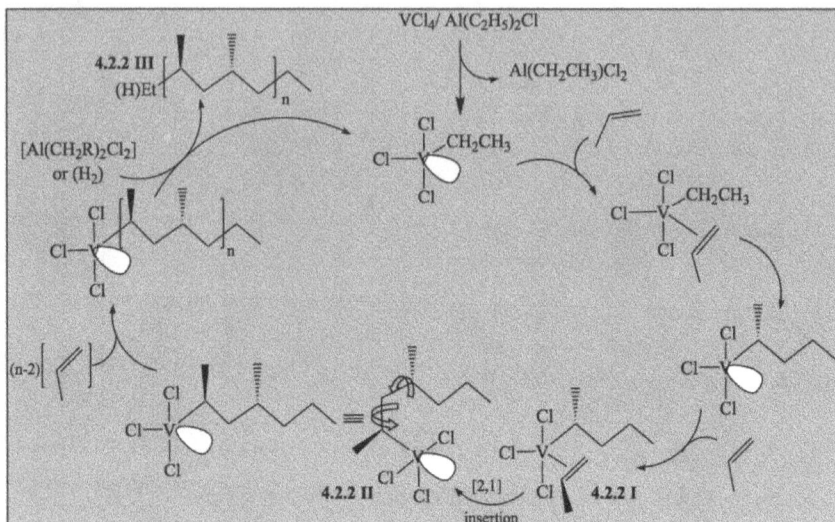

Fig. 4.2.2   V-Based PP Catalysts: Syndiotactic (Ref. 1)

that the secondary alkyl metal bond is produced. This is unexpected since, for most organometallic complexes, the primary metal alkyl is normally produced. In each V alkyl complex, the most stable conformer is with the VCl_3 group and the polymer chain *trans*, as in Figs. 4.2.2 and 4.2.2 I. In each successive addition of propylene, the methyl group assumes a position that is *trans* to the closest methyl group in the polymer chain, as is also shown as 4.2.2 I. After 1,2 insertion occurs, the growing polymer chain on the V alkyl has methyl groups which are oriented as shown as 4.2.2 II. When the polymer chain is rotated around the bonds indicated to produce the most stable configuration of the backbone, the methyl groups are *trans* to one another. This process continues resulting in syndiotactic isomer after chain termination by aluminum alkyl or H_2.

4.2.3 *Ti-Based PP Catalysts: Isotactic*

On the other hand, $TiCl_4/AlEt_3$ catalysts undergo 1,2 migratory insertion:

$$Ti{-}R + CH_2 = CHCH_3 \rightarrow Ti{-}CH_2CH(CH_3){-}R, \qquad (4.2.3)$$

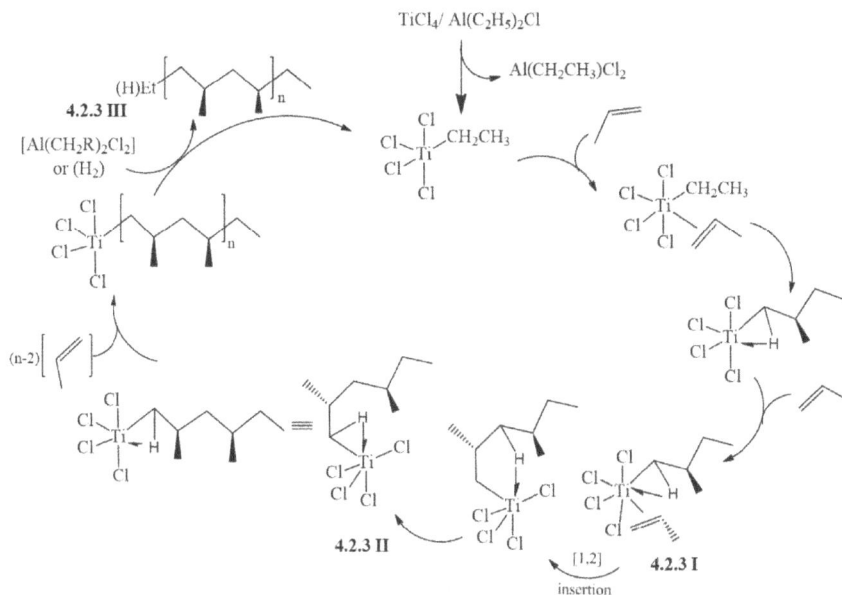

Fig. 4.2.3   Ti-Based PP Catalysts: Isotactic (Refs. 1, 4)

By the same logic as discussed in Sec. 4.2.2, the corresponding intermediates in Figs 4.2.3 I and 4.2.3 II produce isotactic PP (4.2.3 III), as shown in Fig. 4.2.3. There is good evidence that the olefin insertion is facillitated by the α- and β-agostic interactions shown in Fig. 4.2.3 (Ref. 4).

4.2.4 *Ligand Effect: Constrained Geometry Metallocenes*

The formation of atatic versus isotactic isomers in Zr-based metallocene catalysts is an illustration of how ligands, which impart a constrained geometry to the catalytic metal site, can control selectivity (Fig. 4.2.4). The constrained catalytic site of the rac-Cp*$_2$ZrCl$_2$ forms the desired iso-tactic PP, while the un-bridged Cp$_2$Zr(CH$_3$)$_2$ forms the atactic isomer. In this case, the stereoselectivity is controlled by the geometry of the cata-lytic site itself (i.e., *site control*), not by the most favored geometry of the chain end (i.e., *chain end control*), which is operative in the Ziegler–Natta and V-based polypropylene catalysts (Secs. 4.2.2 and 4.2.3). This site-controlled mechanism is illustrated in Fig. 4.2.5.

Ref: Moulijn

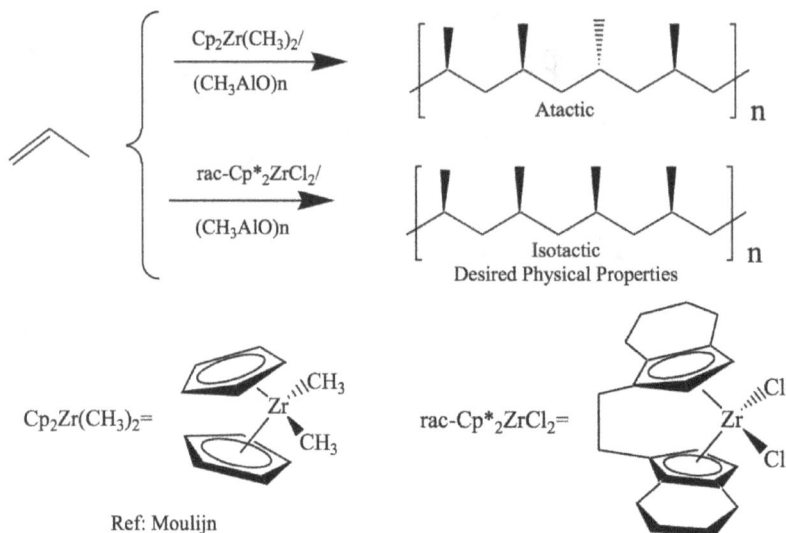

Fig. 4.2.4   Ligand Effect: Constrained Geometry Metallocenes (Ref. 1)

4.2.5 *Isotactic from Cp* Catalyst*

The active Zr species formed (Fig. 4.2.5) coordinates propylene (Fig. 4.2.5 I) and undergoes the expected 1,2 insertion to form Fig. 4.2.5 II. Chirality is then introduced by the coordination of a second propylene to produce Fig. 4.2.5 III, which has several energetically non-equivalent alternatives for migratory insertion of the second and subsequent propylene molecules.

If the plane of the page contains the Zr atom, with the blue indenyl group above and the red one below the plane of the page, the most favored orientation for the methyl groups Fig. 4.2.5 III is away from the closest indanyl group, i.e., upward for the metal alkyl and downward for the methyl group in the coordinated propylene. The result of a 1,2-migratory insertion in this configuration is shown in Fig. 4.2.5 IV. This process repeated several times gives Fig. 4.2.5 V with the stereo chemistry of the growing polymer chain as shown, and subsequently polypropylene Fig. 4.2.5 VI. Rotation around C–C bonds in the backbone gives the most stable conformer as shown, which is clearly the isotactic isomer. It is now believed that an a-agnostic C–H–Zr bond in the metal-alkyl intermediates

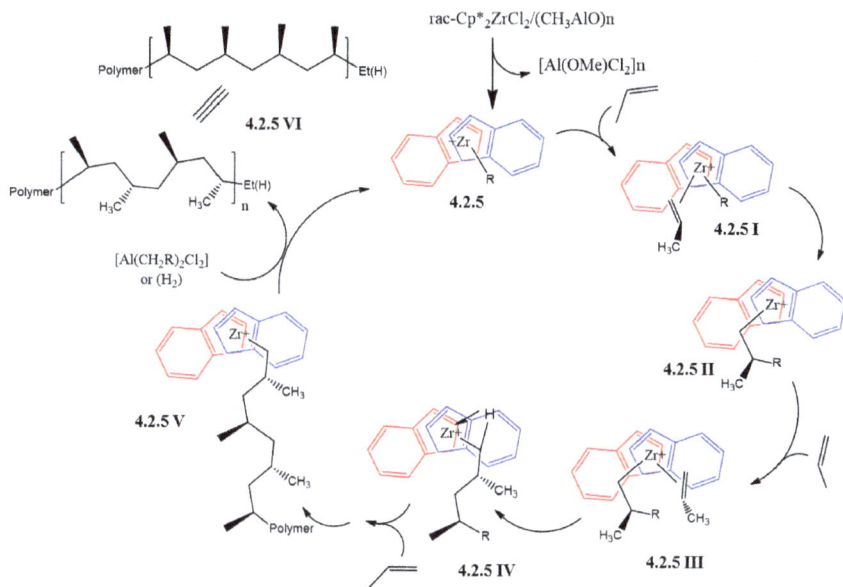

Fig. 4.2.5 **Isotactic from Cp* Catalyst** (Ref. 1)

(as shown in *Structure* Fig. 4.3.5 IV) is a critical factor in controlling the relative orientation of the growing polymer chain in the least sterically-hindered position and directing the selectivity of the insertion of propylene units monomer into the most facially-favorable position (Ref. 4).

4.2.6 *Non-Coordinating Anions*

The use of a *non-coordinating anion*, that is, one which is only weakly associated with the corresponding cation, is important for achieving the desired activity and selectivity for several classes of polymerization catalysts, where the active catalyst center is a cationic species.

4.2.6.1 Activation of catalytic cations

The cationic catalyst center is more active when it is only weakly associated with its counterion. Two examples of this are metallocene polyolefin catalysts and cationic polymerization salt catalysts (Fig. 4.2.6.1). In the

Non-Coordinating Anions Enhance Activity of
Cationic Catalytic Species:

- Metallocene Polyolefin Catalysts:

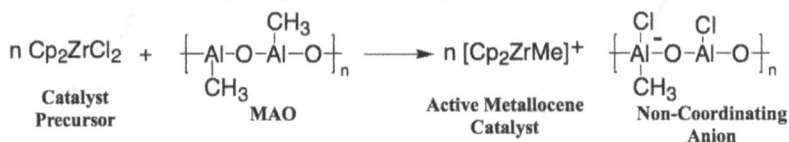

$$\text{n } Cp_2ZrCl_2 + \left.\begin{array}{c} CH_3 \\ | \\ Al-O-Al-O \\ | \\ CH_3 \end{array}\right]_n \longrightarrow \text{n } [Cp_2ZrMe]^+ \left.\begin{array}{c} Cl \quad Cl \\ | \quad | \\ Al-O-Al-O \\ | \\ CH_3 \end{array}\right]_n$$

| Catalyst Precursor | MAO | Active Metallocene Catalyst | Non-Coordinating Anion |

- Cationic Salt Catalysts:

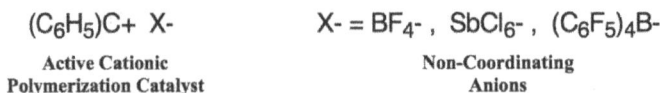

$$(C_6H_5)C+ \; X- \qquad\qquad X- = BF_4- , \; SbCl_6- , \; (C_6F_5)_4B-$$

| Active Cationic Polymerization Catalyst | | Non-Coordinating Anions |

Fig. 4.2.6.1   Non-Coordinating Anion Activation of Catalytic Cations (Refs. 1, 30, 31)

case of the metallocene catalysts, the active Zr^+ metallocene catalayst is formed by reaction of the Cp_2ZrCl_2 catalyst precursor with MAO. MAO is not only a good methyl donor to the Zr center, but it also forms the non-coordinating dichloro-MAO anion, which enhances the reactivity of the Zr^+ center toward olefin polymerization. Likewise, carbon-centered cationic catalysts are most active when a non-coordinating anion is used. An example is trityl tetra(pentafluorophenyl) borate $(Ph_3C^+)[(C_6F_5)_4B^-]$, where (Ph_3C^+) is the active cationic center and $[(C_6F_5)_4B^-]$ is the non-coordinating anion.

4.2.6.2 Effect on reaction path

An example of a catalyst which utilizes a non-coordinating anion is the cationic polymerization of isobutylene (Fig. 4.2.6.2). BF_3 is used as a catalyst for this reaction with the addition of alcohol (ROH) as a proton donor. The reaction of BF_3 and ROH forms an active proton H^+ with the non-coordinating anion, $BF_3(OR)^-$. Beause this anion is only weakly coordinating, the proton is very active and reacts with isobutylene to form the $[(CH_3)_3C^+]/BF_3(OR)^-]$ anion pair. As with the $[H^+][BF_3(OH)^-]$ ion pair, in the $[(CH_3)C^+][BF_3(OH)^-]$ ion pair, $BF_3(OH)^-$ is only weakly associated with the tert-butyl cation $[(CH_3)_3C^+]$, which is believed to

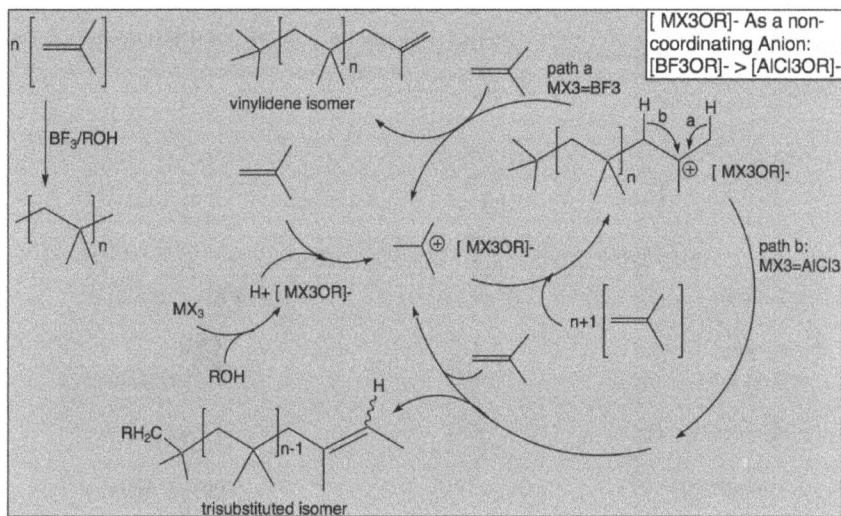

Fig. 4.2.6.2   Non-coordinating Anions — Effect on Reaction Path (Ref. 16, reproduced with permission)

enhance its reactivity toward reaction with more isobutylene monomer, and subsequently the loss of a proton to form the vinylidene isomer of polyisobutylene. Other catalysts such as $AlCl_3/H_2O$ form the $[H^+]$ $[AlCl_3(OH)^-]$ ion pair, where the anion is more coordinating than $BF_3(OH)^-$. This results in a more coordinated $[(CH_3)C^+][AlCl_3(OH)^-]$ ion pair and subsequently, the corresponding $[R^+(AlCl_3OH)^-]$ ion pair after reaction with more polyisobutylene. The coordination of the $[AlCl_3(OH)^-]$ anion slows deprotonation, which enhances rearrangement relative to proton loss, leading to the tri-substituted isomer. Thus, the more reactive vinylidene isomer is formed by catalysts which form non-coordinating anions, including BF_3/ROH and $H_{0.5}(NH_4)_{2.5}PW_{12}O_{40}$.

4.2.7 *Effect of Solvents: Butadiene Example*

Another important means of controlling polymerization reactions is the choice of solvent. For example, in the anionic polymerization of butadiene (Fig. 4.2.7), four major polybutadienes are possible: two 1,4-addition products (*cis* and *trans*), and two 1,2-addition products (syndiotactic and

$$n \ / \!\!/ \backslash \xrightarrow[\text{solvent}]{\text{RLi}} [\text{-CH}_2\text{CH=CHCH}_2\text{-}]n \ + \ [\text{-CH}_2\text{CH-(CH=CH}_2)]n$$

non-polar solvents   polar solvents
favor 1,4-BD   favor 1,2-BD

1,4-cis	1,4-trans	1,2 isotactic n	1,2 syndiotactic n

	ratio affects crystallinity (Tg)		+ 1,2-at* (generally undesired – oxidative instability)
n-hexane	35%	57%	8% (*atactic)
n-hexane/ diethyl ether	29%	50%	21% (*atactic)
n-hexane/THF	25%	40%	35% (*atactic)
heptane/DIPIP*	<1%	<1%	99% (*atactic)

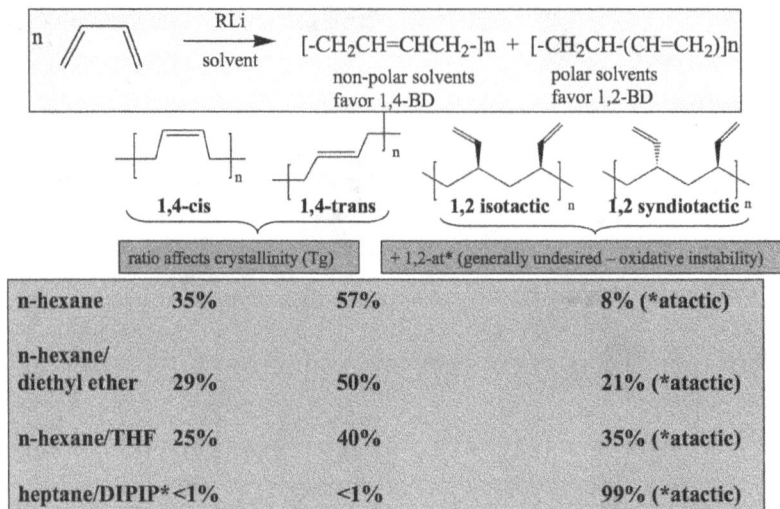

Fig. 4.2.7   Effect of Solvents: Butadiene Example (Refs. 25, 27–29)

isotactic). With alkyl lithium catalyst, non-polar solvents favor the linear 1,4 products, while polar ones favor the branched 1,2-addition products. The polarity of the solvents increases in the table from top to bottom. The 1,4 products are generally desired because they are more oxidatively stable than the 1,2 products.

4.2.8 Ziegler–Natta Catalyst Synthesis: Particle Microreactor

For heterogeneous polymerization catalysts, the catalyst can be used as a particle microreactor, in which the polymer grows around the catalyst particle (Fig. 4.2.8). An example of this is Ziegler–Natta olefin polymerization using $TiCl_3/AlEt_2Cl/MgCl_2$ catalyst. In this system, $AlEt_2Cl$. is first adsorbed onto the spherical $MgCl_2$ support. $TiCl_3$ is then adsorbed onto the resulting material, which converts the $AlEt_2Cl$ layer to a layer of the activated Ti alkyl species (as discussed in 4.1.9) supported on the spherical $MgCl_2$. In the presence of olefin, polymerization occurs, catalyzed by the Ti-alkyl, forming a thick layer of polymer over the catalyst particle.

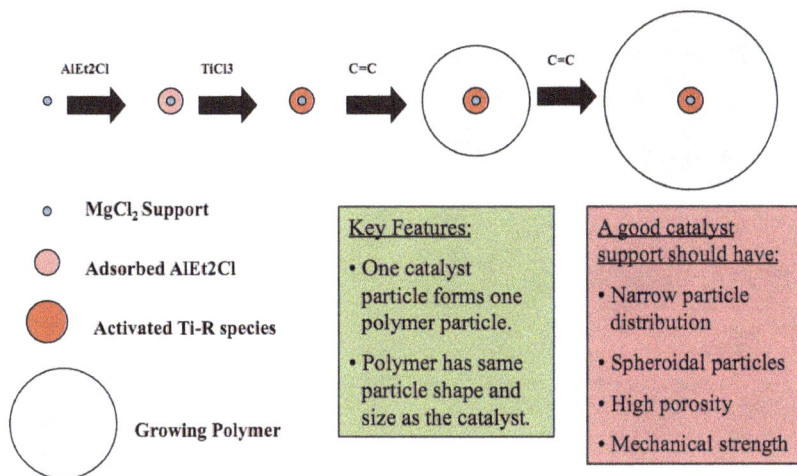

Key Features:
- One catalyst particle forms one polymer particle.
- Polymer has same particle shape and size as the catalyst.

A good catalyst support should have:
- Narrow particle distribution
- Spheroidal particles
- High porosity
- Mechanical strength

Fig. 4.2.8   Z–N Catalyst Synthesis: Particle "Microreactor" (Refs. 1, 30, 31)

By this mechanism, it can be seen that one catalyst particle forms one polymer particle and that the polymer has the same shape and size (initially) as the catalyst. In the final polymer, the catalyst makes up only a very small percent of the total catalyst mass. In order for this method to work, a good catalyst support must have a narrow distribution of spherical particles, of high porosity and mechanical strength.

4.2.9 *Single versus Multiple Site Polyolefin Catalysts*

As we have discussed, Ziegler–Natta polyolefin catalysts are heterogeneous, and contain a distribution of active sites. More recently, catalysts have been discovered that are based on metallocenes, which are based on cyclopentadienyl complexes of rare earth metals, for example $[ZrCp_2CH_3]^+$. A comparison of these systems is shown in Fig. 4.2.9.

Ziegler–Natta catalysts are based on formation of Ti alkyls from reaction of $TiCl_3$ with AlR_2Cl on a catalyst support such as $MgCl_2$ (see 4.2.8). In the solid state, these precursors are present in several different environments, giving rise to a multiple-site, heterogeneous system. Since each of these sites has unique propagation, chain transfer and termination steps, the compositions formed can be broad and difficult to

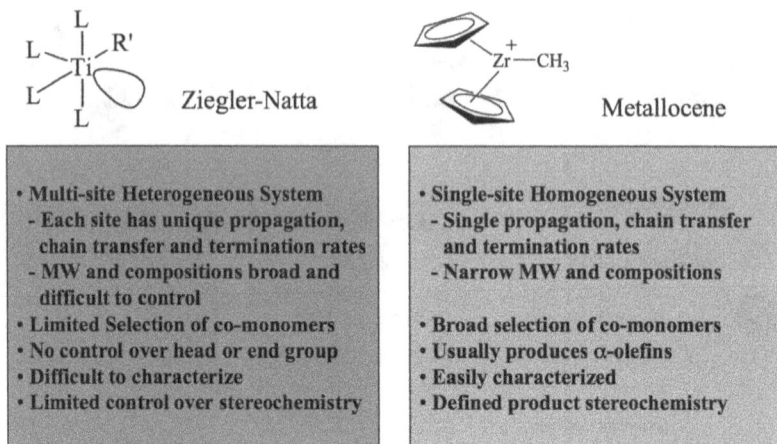

Fig. 4.2.9   Single Versus Multiple Site Polyolefin Catalysts (Refs. 1, 30, 31)

control. In addition, there is a limited selection of *comonomers* (a second olefin monomer which, when added to the polymerization reaction can be incorporated into the primary polymer chain), usually to olefins which are similar to the primary monomer. Furthermore, there is little or no control over the head or end group which is formed. For example, an olefin end group results from chain transfer, whereas a saturated end group forms as a result of chain termination by H_2 or a metal alkyl (see 4.1.9). In addition, these are difficult to characterize and there is only limited control over stereochemistry (see 4.2.1).

Metallocene systems, on the other hand contain only a single active metal site, in this case Zr. This site has a single propagation, chain transfer and termination rates, which gives rise to a narrow molecular weight distribution and composition. Furthermore, this catalyst system can accommodate a broad selection of comonomers, including ones with certain functional groups other than an alkyl group, on the olefin (for example — an ester group, CO_2R, as in acrylates). These catalysts usually produce polymers with olefin end groups by a chain transfer-type mechanism. And since they are pure molecular species, which are used in soluble form, they are easily characterized They can also be designed by organometallic methods of ligand substitution to control steric and electronic effects to produce polymers of defined product stereochemistry (Sec. 4.2.4.)

4.3 Major Industrial Processes

In 4.3 is discussed some of the major commercial catalytic processes by which industrial polymers are made. Features among these are the polyolefins polyethylene, polypropylene and olefin copolymers by Ziegler–Natta and metal oxide catalysts, olefin oligomers by the Shell Ni catalyst, free radical emulsion processes for styrene-butadiene rubber, polyethoxylates by anionic catalysts, and cationic catalysts for polyisobutylene. Following this section is a discussion of a future trend in polyolefin catalysts — late transition metal catalyst systems for production of branched polyethylene.

4.3.1 *Polyethylene and Isotactic Polypropylene:* $TiCl_4/AlEt_2Cl/MgCl_2$

Modern commercial processes for polyethylene and the preferred isotactic form of polypropylene use the modified Ziegler–Natta system: $TiCl_4/AlEt_2Cl_2/MgCl_2$. The mechanism (Fig. 4.2.3) follows the same 1,2-insertion scheme was discussed in 4.2.3. The use of highly purified $MgCl_2$ support along with the use of oxygen (Lewis base) donors account for much of the improvement in activity and selectivity to isotactic isomer for current state-of-the-art Ziegler–Natta catalysts. This is discussed in more detail in Fig. 4.3.1.1.

4.3.1.1 Ziegler–Natta catalyst synthesis particle "Microreactor"

The positive effect of O^- (Lewis base) donors on the properties of the catalyst support can be understood in the context of the particle microreactor system, as was introduced in 4.2.8. In this scheme, the $MgCl_2$ support is first treated with an oxygen donor molecule. The donor molecules improve catalytic performance in two ways: (i) by enhancing the interaction between spherical support and the Al and Ti components, leading to improve reactivity of active sites and (ii) by interaction with and blocking the more exposed sites, which are believed to be less stereospecific to the isotactic form. Specific donor molecules include phthalate esters like di-n-butyl and di-isobutyl phthalate, and 1,3-diethers $(R_2C(CH_2OR)_2$. (Refs. 25–29, 30, 31, 34).

The donor system when used with the Ziegler–Natta cataylst discussed in Sec. 4.2.8 have improved selectivity to isotactic polypropylene. Examples of specific donors and their effects on polymerization performance are discussed in Sec. 4.3.1.2.

4.3.1.2 Polyethylene and isotactic polypropylene: donor molecule improvement

The improvement in activity and selectivity to isotactic polypropylene isomer resulting from the use of O-donor (Lewis base) molecules is shown in Table 4.3.1.2. Activity for PP and selectivity to isotactic PP and for the di-ether donor molecule (1) and more traditional di-isobutyl phthalate (2) phenyl triethoxysilane (3) donor system versus the same $MgCl_2$/$TiCl_3$/$AlEt_3$ Ziegler–Natta catalyst with no donor are shown in Table 4.3.1.2.

There has been much research into the mechanism of the action of Lewis base O-donors (Ref. 34). While no single explanation accounts for all the results, there is general consensus that binding of donors to catalyst sites can have a profound effect on activity and selectivity to isotactic isomer.

4.3.2 *Ethylene/Propylene Copolymers (EPC)*

4.3.2.1 General

Incorporation of ethylene and propylene into the same polymer by *copolymerization* of the two monomers is an important means of tailoring polymer properties to achieve the optimal balance of *crystalline* (from polyethylene) and *amorphous* regions (from polypropylene). Crystalline materials have long range order (which is why they produce an X-ray diffraction pattern), while amorphous ones lack such long range ordering of atoms. The amorphous polypropylene structure is generally preferred

1a   1c

1b   2   3

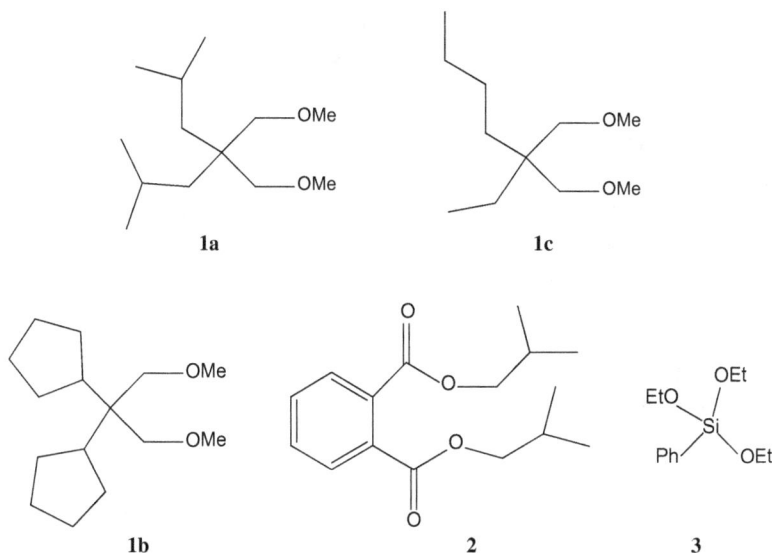

Table 4.3.1.2: Activity and Isotactic Selectivity of $MgCl_2$–$TiCl_4$–Donor/ $AlEt_3$ Catalyst for Propylene Polymerization (70°C, 2 hr) (Ref. 34)

Donor	Activity, kg PP/g cat	Selectivity, % Isotactic
1a	80	98%
1b	50	98%
1c	35	91%
2 + 3	40	98%
None	24	42%

because of its higher tensile strength, improved stress crack resistance, greater rigidity and higher melting point, important for thermoplastic applications.

But polyethylene is much cheaper. Thus, it is desirable to add enough propylene to the copolymer to achieve an amorphous state. In fact, incorporation of as little as 25–40% of propylene into a copolymer of ethylene and propylene gives an amorphous polymer, but copolymers with up to 70% propylene are

4.3.2.1 Ethylene/ Propylene Copolymers

- Incorporation of a second polymer provides a means of tailoring polymer properties
- Incorporation of propylene (amorphous) into polyethylene (crystalline) controls level of crystallinity in the resulting copolymer

- 25–40 mole% incorporation of propylene into polyethylene gives an amorphous polymer
- Ti/Al-based Ziegler–Natta catalysts do not incorporate propylene well because they polymerize ethylene much faster than propylene
- V and Cr-based catalysts are commercially used because monomer reactivities are much closer

commercially available. These copolymers are known as *OCP's* or *Olefin Copolymers*. Ziegler–Natta systems can also be used form OCP's, but V- and Cr-based systems are preferred, because catalysts based on Ti and Al polymerize ethylene much faster than propylene and therefore, are not very effective at incorporating propylene into the polymer.

4.3.2.2 Mechanism

The mechanism for the formation of OCP's using V-based Z–N catalysts is shown in Fig. 4.3.2.2. This is analogous to the mechanism already discussed in 4.2.2, except that both ethylene and propylene are incorporated into the growing polymer. V catalysts produce largely randomly distributed copolymers, which minimizes the long runs of ethylene in the polymer that would give rise to crystalline polymer.

Various other comonomers are used to modify the properties of polyethylene. Pure polyethylene is known as *low density polyethylene* (LDPE). *High density polyethylene* (HDPE) contains 2 mole % of a longer 1-alkene such as 1-octene, which has just enough of the higher olefin to disrupt crystallinity, making the material less brittle and susceptible to stress cracking. Linear low density polyethylene (LLDPE) contains 8–10% of a 1-alkene such as 1-octene, 1-hexene or 1-butene, which has an even greater effect on reducing crystallinity, with the corresponding improvement in properties. As for OCP's, the goal of these materials, is to improve the

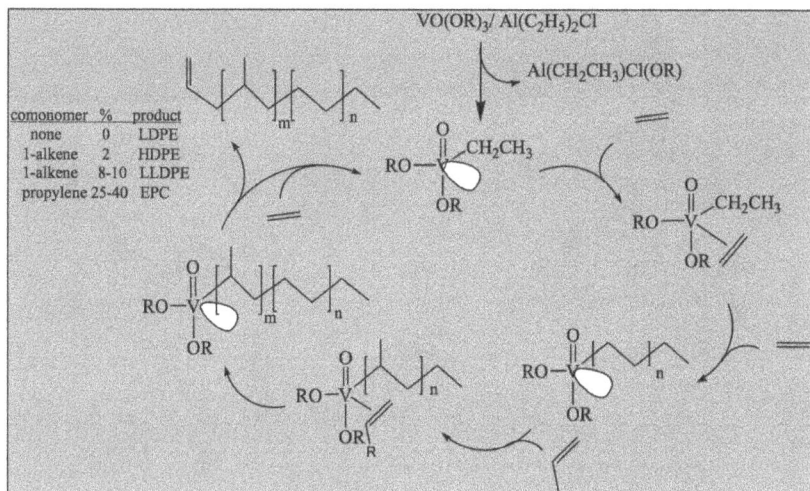

Fig. 4.3.2.2 OCP Mechanism (Ref. 25–29, 30)

performance of polyethylene by introduction of a second monomer, in as small amounts as is necessary, to disruption the perfectly linear structure, thus reducing cyrstallinity and improving its physical and mechanical properties.

4.3.3 *Higher Olefins: Shell Higher Olefin Process (SHOP)*

Higher olefins in the C6–C42 range are used as chemical intermediates for many products, including as comonomers for polyethylene as discussed in 4.3.2. These can be thought of a short polymers or oligomers, and are produced by a process known as the Shell Higher Olefin Process (Fig. 4.3.3). This process uses an organo-Ni catalyst with a bidentate P- and O-containing ligand. The mechanism is very similar to the metallocene polyethylene mechanism (4.1.10), in that it contains ethylene coordination, addition and β-elimination steps to produce polymer with an olefin end group. In this case, the number of ethylene monomers in the product is in the 3–23 range ($M_n = 168$–644), versus many more for metallocenes, which can make polymers in the 100,000–1,000,000 molecular weight range. The Ni system with the P/O ligand has a reduced ratio of

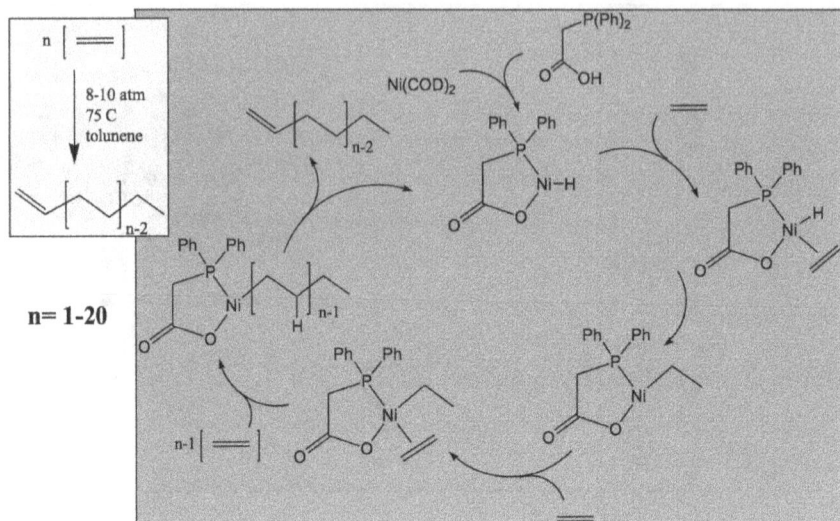

Fig. 4.3.3   SHOP Mechanism (Refs. 1, 18, 30, 35, 36)

propagation (ethylene insertion) to chain transfer (β-elimination) compared to the Zr-metallocene system, giving rise to a shorter chain length for the former.

4.3.4 *Styrene/Butadiene Emulsion Polymerization (SBR)*

4.3.4.1 General

Styrene butadiene emulsion polymerization is an example of a radical-initiated polymerization which occurs in an oil-in-water emulsion (Fig. 4.3.4.1). The monomers (butadiene and styrene) and a chain transfer agent (dodecycl mercaptan to control molecular weight) are emulsified by sodium dodecylbenzene sulfonate (SDS) in an aqueous phase which contains the potassium persulfate initiator. The SDS emulsifier stabilizes the emulsion by formation of a micellular structure in which the polar head (the sulfonate group — SO_3^-) of the emulsifier sticks into the aqueous phase while the hydrocarbon tail is dissolved in the oily monomer phase. This alignment of emulsifier molecules stabilizes the oil phase as small spheres suspended in the aqueous phase, and prevents the coalescence of these oily spheres into larger spheres by the electrostatic and

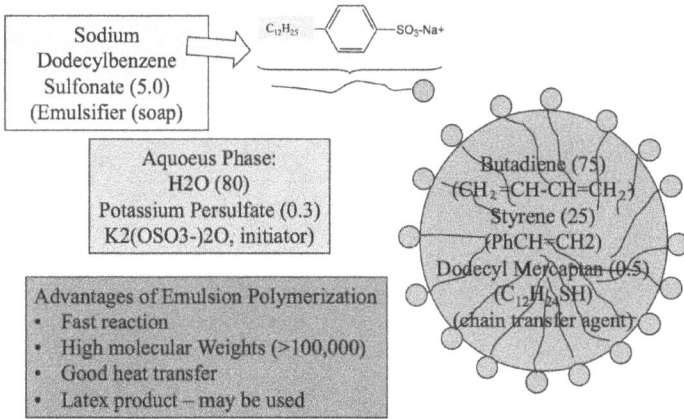

Fig. 4.3.4.1   Styrene/Butadiene Emulsion Polymerization (Refs. 25–29, 30)

steric repulsion between the SDS molecules in two micellular structures. The advantages of this method of polymerization include a rapid reaction rate, high molecular weights (>100,000) and good heat transfer. The product, in which the spherical micelles of monomer are emulsified in the aqueous phase, and have been converted into S–B rubber, is known as a *latex*. In many cases, the latex product can be used in subsequent processing without the need to remove the water. For example, carbon filler can be added and the resulting compounded rubber fabricated into reinforced rubber products, including tires, hoses and belts.

4.3.4.2 SBR mechanism

The mechanism of SBR production (Fig. 4.3.4.2) follows the general scheme of free radical initiated polymerization, as was discussed in 4.1.11 for polystyrene, except that now a second monomer — butadiene — is incorporated into the polymer chain. Initiation occurs by cleavage of O–O bond in the persulfate and addition of the resulting sulfate radicals into the styrene monomer. The resulting carbon-centered radical continues to added styrene monomer, and subsequently butadiene molecules to form the growing polymer radical chain. The mercaptan (RSH) serves as a chain transfer agent, by donation of an H• atom to terminate the chain, producing SBR, andgenerating RS•, which can initiate additional polymerization chains.

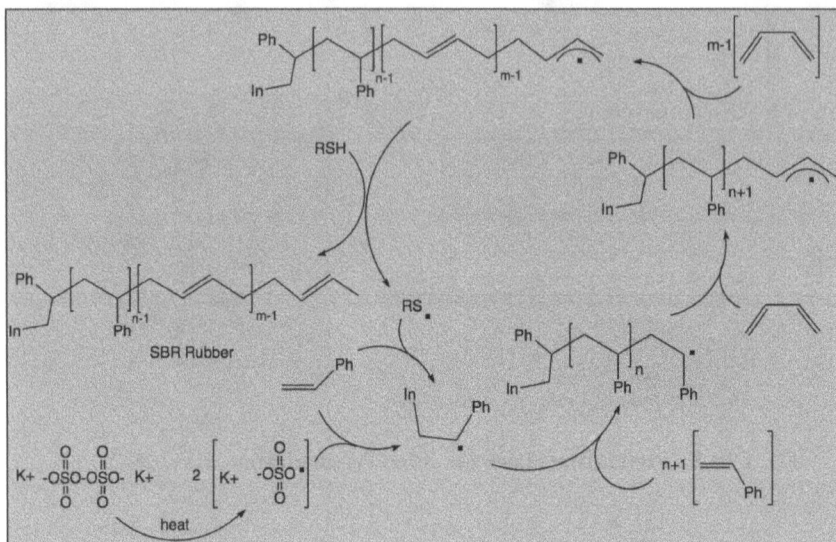

Fig. 4.3.4.2 Styrene/Butadiene (Refs. 25–30)

4.3.5 *Polyethoxylates (Anionic Polymerization)*

One of the most prominent anionic polymerization processes is the production of polyethoxylates from epoxides using KOH as catalyst (Fig. 4.3.5). The reaction proceeds by nucleophilic attack of OH^- on the epoxide ring, followed by reaction of the resulting alkoxide with more epoxide monomer. Chain transfer occurs after about 6–20 monomer units have reacted. Polyethoxylates are very important as industrial non-ionic surfactants, which appears in many consumer products such as fabric softeners and hair conditioners.

4.3.6 *Polyisobutylene*

4.3.6.1 General

One of the best examples of an industrial high volume polymer made by cationic polymerization is polyisobutylene. There are four major end groups that can be formed: vinylidene, β-isomer, tri-and tetra-substituted (Fig. 4.3.6.1). These appear in two types of products: conventional PIB, made from $AlCl_3$ catalysts, which comprised mainly tri- and

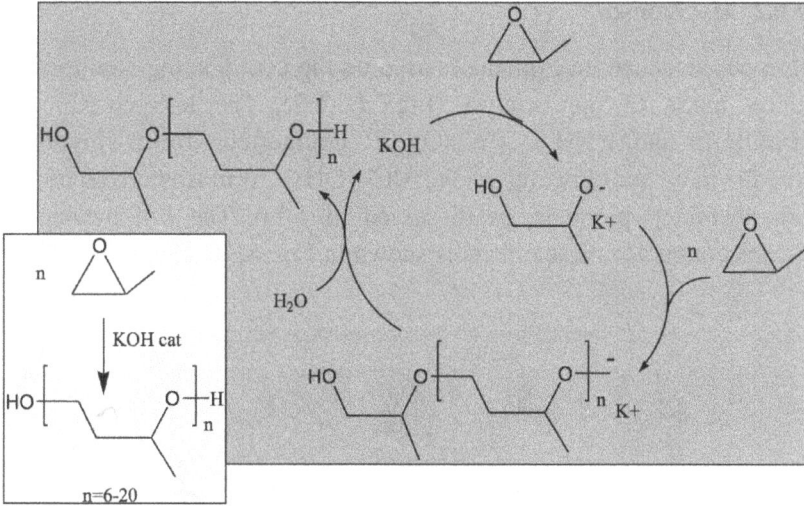

Fig. 4.3.5   Polyethoxylates (Anionic) (Refs. 25–30)

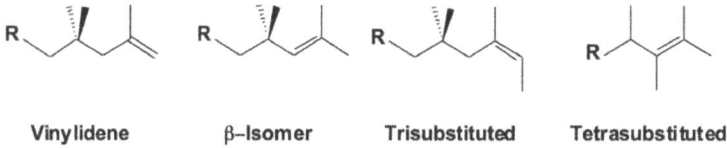

| Vinylidene | β-Isomer | Trisubstituted | Tetrasubstituted |

PIB Type	Catalyst	Mn (Peak)	Mw/Mn	Vinyl	Tri	β-isomer	Tetra
Isomerized End Group	AlCl$_3$	1,000-2,000 (3,000)	2.5-3.5	5%	65%	5%	25%
High Vinylidene	BF$_3$	1,000-5,000 (3,000)	1.3-3.0	70-85%	-	10-20%	2-5%
High Vinylidene	M$_{2.5}$H$_{0.5}$PW$_{12}$O$_{40}$	1,000-3,500 (40,000)	5 - 25	70-85%	-	10-20%	2-5%

Fig. 4.3.6.1   Polyisobutylene (Ref. 16, reproduced with permission)

tetra-substituted forms, and high vinylidene PIB, which is mainly the vinylidene form, with smaller amounts of the β-isomer, made from BF$_3$ and heteropolyacid catalysts.

The former result from isomerization of the end group, while the latter are a product of direct β-elimination from an un-isomerized end group.

4.3.6.2 Mechanism

These products can be explained based on the coordinating nature of the counter anion of the catalyst (Fig. 4.3.6.2). The less coordinating $[BF_3(OH)]^-$ and $[(NH_4)_{2.5}PW_{12}O_{40}]^{-1/2}$ counterions from vinylidene, while the more coordinating anion $[AlCl_3(OH)]^-$ form isomerized tri- and tetra-substituted products, as discussed in 4.2.6. The full mechanism which accounts for these effects is shown in Fig. 4.3.6.2.

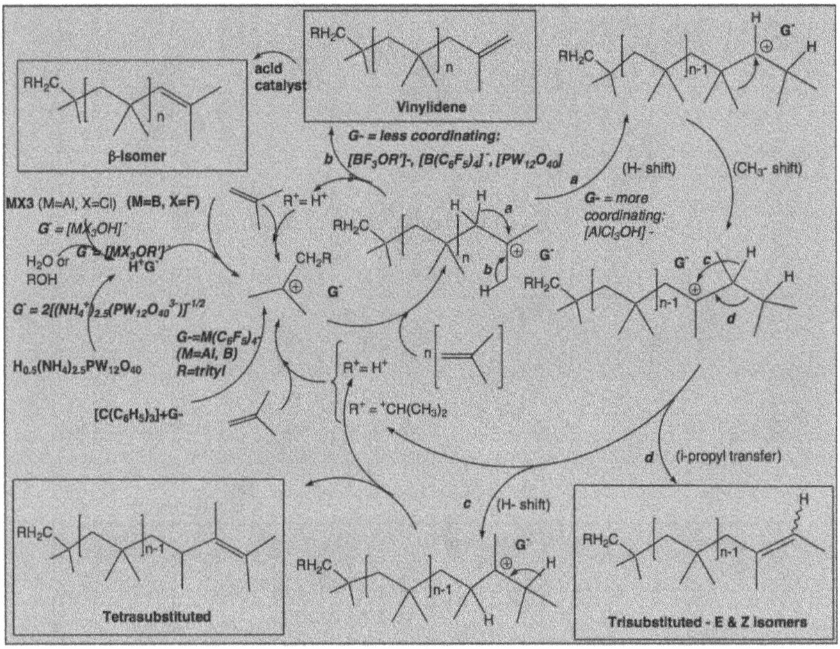

Fig. 4.3.6.2 Mechanism (Ref. 16, reproduced with permission)

4.4 Future Catalysts: Late Transition Metals for Branched Polyethylene

One of the most exciting developments in polymerization catalysis in recent years is the discovery of late transition metal catalysts for production of branched polymers of ethylene (Fig. 4.4).

One of the first class of catalyst systems in this category is the Brookhart catalyst, discovered by Maurice Brookhart and coworkers at

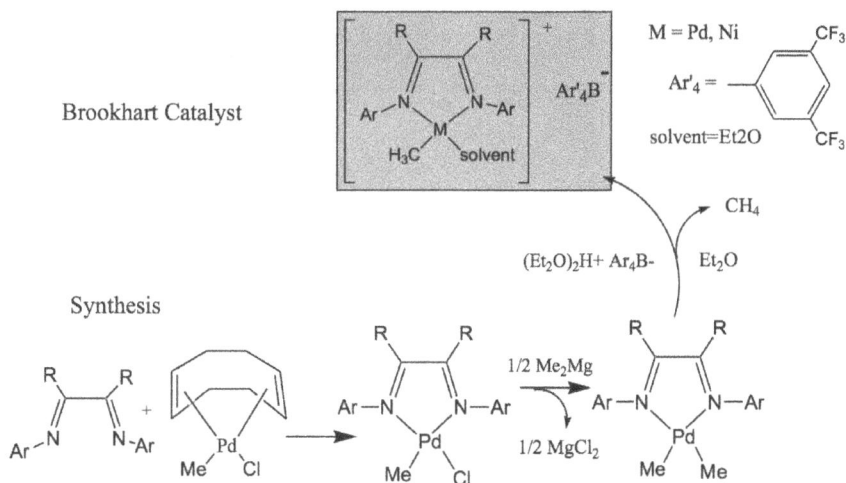

Fig. 4.4 **Future Catalysts: Late TM for Branched PE** (Ref. 37)

the University of North Carolina, Chapel Hill. In one embodiment, like the SHOP catalyst (Sec. 4.3.3), it is an organonickel complex with a bidentate ligand, but in this case, it is a neutral *N, N* di-imine ligand versus the anionic P, O ligand. The central metal can also be another *late transition metal* (i.e., a transition metal toward the right of the periodic table), most notably Ni or Pd. Like the polyisobutylene (4.1.7) and metallocene systems (Sec. 4.1.10), this catalyst works best with a non-coordinating counter anion, in this case a tetra aryl borate $(Ar4B)^-$. The catalyst is synthesized by reaction of the metal cyclooctadiene (COD) methyl chloride complex with a diimine ligand and methyl magnesium. The organomethyl acts as a source of methyl anion (CH_{3-}), displacing the chloride from the Pd, producing the desired Pd (2+) di-imine, dimethyl catalyst complex.

4.4.1 *Branched Polyethylene*

The unique features of this catalyst is that it produces a polymer with alkyl branches, using ethylene as the sole monomer. Of course, using conventional Ziegler–Natta or metallocene polyethylene catalysts, the production of branched polymer would require addition of an alpha-olefin comonomer. A key to the operation of this catalyst is the use of sterically

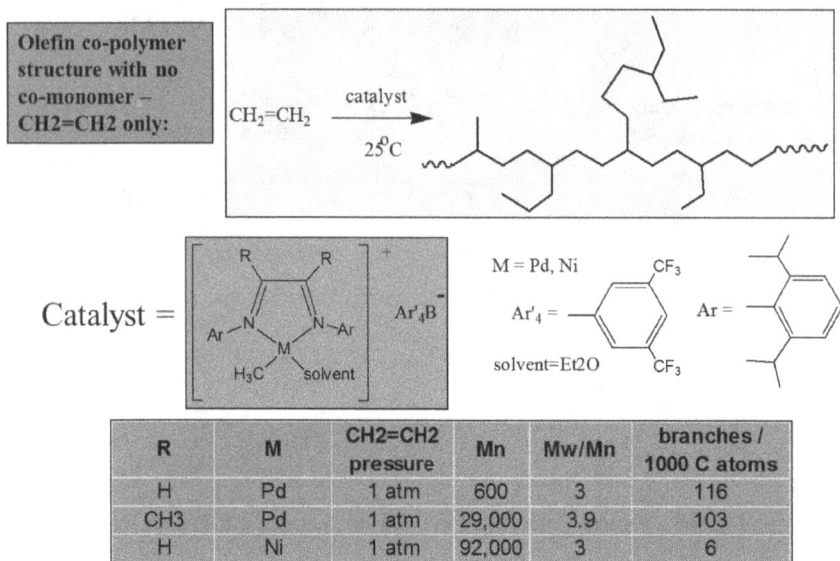

R	M	CH2=CH2 pressure	Mn	Mw/Mn	branches / 1000 C atoms
H	Pd	1 atm	600	3	116
CH3	Pd	1 atm	29,000	3.9	103
H	Ni	1 atm	92,000	3	6

Fig. 4.4.1   Branched PE (Ref. 37)

constrained ligands Ar and Ar′ (as shown in Fig. 4.4.1). The selection of the imine alkyl group (R) and the metal (Ni versus Pd) in the catalyst control the molecular weight (M_n) and level of branching (number of branches per 100 C atoms). Where R=H, low molecular weight $(M_n = 600)$ polymer is formed with 116 branches per 1000 C atoms. The substitution of R with methyl groups creates more steric constrains resulting in a less active site (which requires about an order of magnitude greater H_2 pressure), but in a much higher molecular weight $(M_n = 29,000)$, and slightly less branching (103 branches per 1000 C atoms).

Mechanistically, the incorporation of branches from ethylene alone requires some sort of isomerization and equilibration of ethylene groups in the metal alkyl intermediates. A mechanism consistent with these observations is discussed in 4.4.2.

4.4.2 Branching Mechanism

The key reaction steps in the mechanism are shown in Fig. 4.4.2. The first step involves the displacement of ether (solvent) from the active

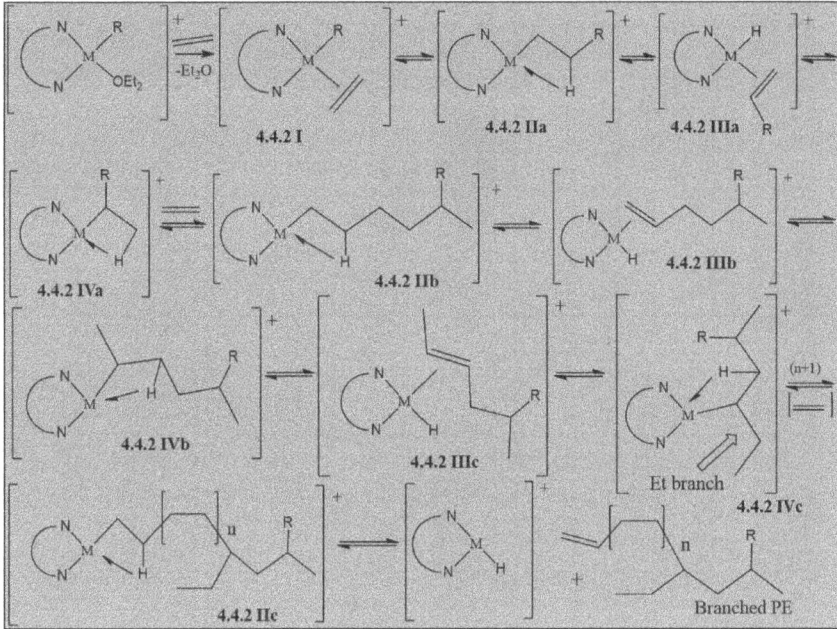

Fig. 4.4.2   Polyethylene Branching Mechanism (Refs. 37–39)

organometallic cation (M = Ni or Pd) by ethylene monomer to form the initial olefin metal alkyl (Fig. 4.4.1I). Ethylene insertion into the β-agostic metal alkyl complex (Fig. 4.4.2II (a)) and β-elimination produces a new metal olefin hydride Fig. 4.4.2III (a). 2,1-olefin addition produces the branched β-agostic metal alkyl Fig. 4.4.2Iv (a). By repetition of these same basic ethylene addition/β-elimination steps, the appearance of methyl (Fig. 4.4.2IV (b)), ethyl (Fig. 4.4.2V (c)) and even longer alkyl branches can be explained in the product, branched polyethylene, which contains an olefin end group resulting from the final β-elimination step.

4.5 Problems

1. Use the gel permeation chromatography (GPC) results shown in the graph below to: select the correct order of polydispersity, from

lowest to highest:

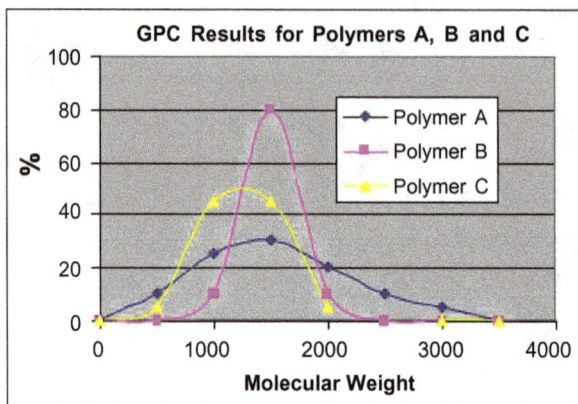

(a) A < B < C,

(b) B < C < A,

(c) C < B < A.

2. Match a polymer from Column A with a catalyst that can be used to produce that polymer from Column B:

A	**B**
(a) Isotactic Polypropylene	(i) BF_3
(b) Polypropylene Oxide	(ii) $TiCl_4/AlEt_2Cl/MgCl_2$
(c) Styrene Butadiene Rubber	(iii) K/toluene
(d) High Vinylidene Polyisobutylene	(iv) $VCl_4/AlEt_3$
(e) Conventional (Trisubstituted) Polyisobutylene	(v) $VO(OR)_3/AlEt_2Cl$
(f) Krayton Rubber	(vi) (1) Pd(COD)MeCl/diphenyl, dimethyl diamine, (2) Me_2Mg
(g) Syndiotactic Polypropylene	(vii) $Ni(COD)_2$
(h) Branched Polyethylene	(viii) AIBN
(i) C16H32 (α-olefin)	(ix) $AlCl_3$
(j) OCP	(x) KOH

(a) _____   (c) _____   (e) _____   (g) _____   (i) _____
(b) _____   (d) _____   (f) _____   (h) _____   (j) _____

3. Match each of the polymerization mechanisms (Column A) with the corresponding catalyst (Column B):

A	**B**
(a) Free Radical	(i) $AlCl_3$
(b) Anionic	(ii) Cp_2ZrCl_2/MAO
(c) Cationic	(iii) $TiCl_3/AlEt2Cl/MgCl_2$
(d) Coordination	(iv) $K_2(OSO^3{}^-)_2/H_2O/C_{12}H_{24}SH/SDS$
(e) Emulsion Free Radical	(v) DPB
(f) Particle "Microreactor"	(vi) KNH_2

(a) _____   (c) _____   (e) _____
(b) _____   (d) _____   (f) _____

4. Match each of the catalyst activation steps (Column A) with the corresponding catalysts (Column B):

A	**B**
(a) Free Radical	(i) In-In $\rightarrow$ 2In•
(b) Anionic	(ii) $Na + NH_3 \rightarrow NaNH_2 + 1/2H_2$
(c) Cationic	(iii) $AlCl_3 + H_2O \rightarrow [AlCl_3(OH)]- H+$
(d) Ziegler-Natta	(iv) $MCl_4 + AlEt_2Cl \rightarrow [MCl_3Et] + AlEtCl_2$
(e) Metallocene	(v) $nCp_2ZrCl_2 + (CH_3AlO)n \rightarrow n[Cp_2ZrCH_3]+ + n[CH_3Al(Cl)-O-Al(Cl)-O]-$

(a) _____   (c) _____   (e) _____
(b) _____   (d) _____

5. Match each of the olefin activation steps (Column A) with the corresponding catalysts (Column B):

(a) ——— (c) ——— (e) ———

A	**B**
(a) Free Radical	(i) In• + RCH=CHR → [InCH(R)-CHR] •
(b) Ziegler–Natta	(ii) CH_2=CH–CH=CH_2 + K• → [CH_2–CH–CH–CH_2] • K+
(c) Metallocene	(iii) H+ + RCH=CHR → [RCH_2–CHR]+
(d) Anionic	(iv) [$LMCl_3$Et (CH_2=CH_2)] → $LMCl_3$ (CH_2CH_2Et)
(e) Cationic	(v) [Cp_2ZrCH$_3$(CH_2=CH_2)]+ → [Cp_2Zr ($CH_2CH_2CH_3$]+

(b) ——— (d) ———

6. Match each of the descriptions in Column A with all of the corresponding mechanistic reaction steps in Column B which apply:

A

(a) Activation of monomer toward polymerization.
(b) The successive addition of monomer units to the growing polymer chain.
(c) Polymer-forming process which involves the transfer of H+, H• or H⁻ from the reactive end of a growing polymer chain to monomer.
(d) Generally forms a polymer with a saturated (sp3 C) polymer end group.
(e) Polymer-forming process which involves the transfer of a H or reactive group to the reactive end of a growing polymer chain.
(f) Generally forms an unsaturated (sp2 C) polymer end group.
(g) Product forming process that does not regenerate the active polymerization intermediate.
(h) Product forming process that regenerates the active polymerization intermediate.

(i) The formation of an olefin-metal hydride complex from a metal alkyl with at least one H on the carbon once removed from the metal-C bond.

(j) The formation of a metal alkyl with more C atoms in the alkyl group from a metal alkyl olefin complex.

B

(i) β-elimination
(ii) Termination
(iii) Propagation
(iv) Chain transfer
(v) Initiation
(vi) Insertion

(a) ———   (c) ———   (e) ———   (g) ———   (i) ———

(b) ———   (d) ———   (f) ———   (h) ———   (j) ———

7. Write a chemical reaction representing each of the processes listed in Column B of Question 6.

(i)

(ii)

(iii)

(iv)

(v)

(vi)

4.6 Answers to Problems

1.

   (b) B < C < A

2.

   (a) <u>ii</u>   (c) <u>viii</u> (e) <u>ix</u>   (g) <u>iv</u>   (i) <u>vii</u>
   (b) <u>x</u>   (d) <u>i</u>   (f) <u>iii</u> (h) <u>vi</u>   (j) <u>v</u>

3.

   (a) <u>v</u>   (c) <u>i</u>   (e) <u>iv</u>
   (b) <u>vi</u>   (d) <u>ii</u>   (f) <u>iii</u>

4.

   (a) <u>i</u>   (c) <u>iii</u>   (e) <u>v</u>
   (b) <u>ii</u>   (d) <u>iv</u>

5.

   (a) <u>i</u>   (c) <u>v</u>   (e) <u>iii</u>
   (b) <u>iv</u>   (d) <u>ii</u>

6.

   (a) <u>v</u>   (c) <u>iv</u>   (e) <u>ii</u>   (g) <u>ii</u>   (i) <u>i</u>
   (b) <u>iii</u>   (d) <u>ii</u>   (f) <u>iv</u>   (h) <u>iv</u>   (j) <u>vi</u>

7. Write a chemical reaction representing each of the processes listed in
   Column B of Question 6.

(i)

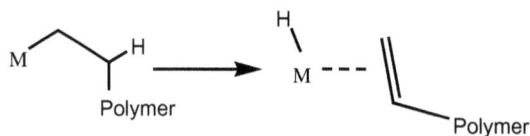

(ii)

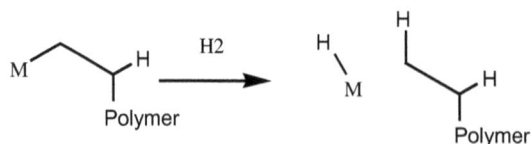

(iii)

(iv)

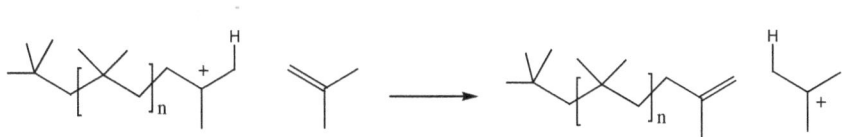

(v)

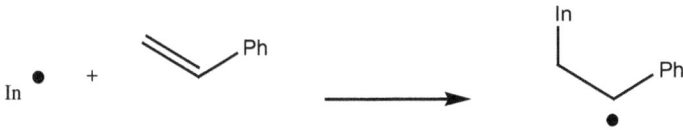

(vi)

5

Reduction/Hydrogenation Catalysis

Catalysts which operate in a reducing atmosphere are used for both chemicals and fuels production. Hydrogen, the simplest and least expensive reductant, is involved as either the reactant or product in these reactions.

The discussion of reduction/hydrogenation processes begins with a look at the elements that dominate these catalysts and the thermodynamic limitations of these reactions. Important mechanistic reaction steps include H_2 activation by metals, CO insertion, reactions of *synthesis gas* $(CO + H_2)$, and CO dissociation. Also discussed are the pathways for the production of the characteristic (Schultz–Flory) distribution of hydrocarbons formed from synthesis gas, and the reductive removal of S as H_2S from organic molecules by a process known as *hydrosulfurization* or HDS.

Important elements of catalyst design that will be discussed include the selection of metals, ligands, promoters and poisons. Some of the major industrial reduction/hydrogenation processes by which high volume chemicals and fuels are produced include refinery *hydrocracking* and *hydrotreating* (*hydrogenolysis* reactions), *hydroformulation* to produce *oxoalcohols*, *carbonylation* of methanol to acetic acid, catalytic refinery *reforming* and ammonia synthesis by the *Haber–Bosch process*.

5.1 Concepts

Table 5.1.1.1: Reduction Catalyst Periodic Table

1a	2a	3b	4b	5b	6b	7b	8			1b	2b	3a	4a	5a	6a	7a	0
H	He																
Li	Be											B	C	N	O	F	Ne
Na	Mg											Al	Si	P	S	Cl	Ar
K	Ca	Sc	Ti	V	Cr	Mn	Fe	Co	Ni	Cu	Zn	Ga	Ge	As	Se	Br	Kr
Rb	Sr	Y	Zr	Nb	Mo	Te	Ru	Rh	Pd	Ag	Cd	In	Sn	Sb	Te	I	Xe
Cs	Ba	La*	Hf	Ta	W	Re	Os	Ir	Pt	Au	Hg	Tl	Pb	Bi	Po	At	Rn
Fr	Ra	Ac**															
*Lanthanides	Ce	Pr	Nd	Pm	Sm	Eu	Gd	Tb	Dy	Ho	Er	Tm	Y	Lu			
**Actinides	Th	Pa	U	Np	Pu	Am	Cm	Bk	Cf	Es	Fm	Md	No	Lw			

Table 5.1.1.2: Reduction Catalyst Periodic Table

5.1.1 *Reduction Catalysis Periodic Table*

Elements involved in reduction and hydrogenation catalysts include the transition elements in low oxidation states, and of course H from dissociation of H_2O.

Prominent among these elements are the noble metals Pt, Pd, Rh and Ir, and the base metals Fe, Co, Ni, Cu and Zn.

5.1.2 *Thermodynamics*

As for polymerization catalysis, the thermodynamically favored products depend on the partial pressure of hydrogen. At low partial pressures, coke formation becomes thermodynamically favored, whereas at high partial pressures, hydrogenolysis to methane and lower hydrocarbons is favored.

An additional thermodynamic concern involves the reactions of CO and hydrocarbons at low H_2 pressures. The most favorable reaction of CO is carbon formation (−38 Kcal/mole), which is also the least desirable, for example, with respect to reaction with water to form CO_2 and H_2 (known as the *water–gas shift reaction*) (−10 Kcal/mole). The production of H_2 and CO (*synthesis gas*) from *steam reforming* of hydrocarbons is actually an endothermic process, which requires a heat source and high temperatures to accomplish.

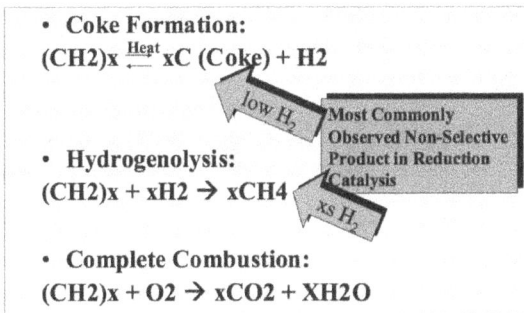

• **Coke Formation:**
$(CH2)x \xrightarrow{Heat} xC$ (Coke) + H2

• **Hydrogenolysis:**
$(CH2)x + xH2 \rightarrow xCH4$

Most Commonly Observed Non-Selective Product in Reduction Catalysis

low H_2

xs H_2

• **Complete Combustion:**
$(CH2)x + O2 \rightarrow xCO2 + XH2O$

Fig. 5.1.2   Thermodynamics

Table 5.1.2: Thermodynamics (Ref. 18)

	ΔH	
$2\,CO \rightleftharpoons C + CO_2$	−38 Kcal/mol,	(5.1.2.1)
$CO + H2O \rightleftharpoons CO_2 + H_2$	−10 Kcal/mol,	(5.1.2.2)
$(-CH2-)n + H2O \rightleftharpoons nCO + 2nH_2$	+36 Kcal/mol.	(5.1.2.3)

5.1.3 *H₂ Activation by Metals/Olefin Hydrogenation*

A key step in the catalytic hydrogenation process is the activation of H_2 by metals, such as it occurs in the hydrogenation of olefins, as shown in Fig. 5.1.3. The H_2 activation step involves its dissociation on a reduced metal surface to form two M–H bonds. Olefin adsorption on adjacent metal sites and reductive elimination produces the corresponding paraffin.

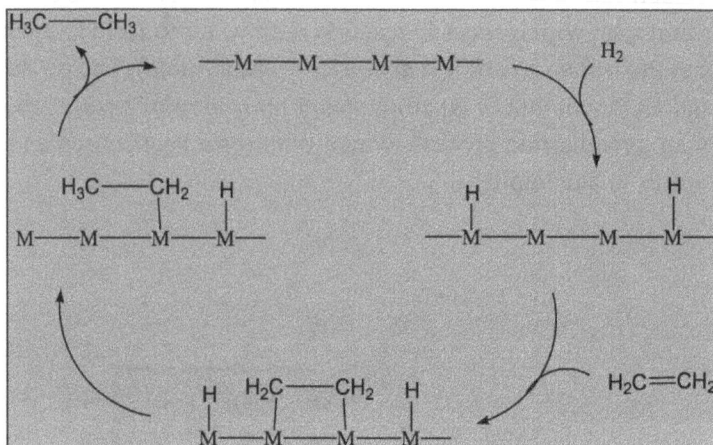

Fig. 5.1.3   H₂ Activation by Metals/Olefin Hydrogenation

5.1.4 *Hydroformylation: CO Insertion*

Another key reaction step in hydrogenation catalysis is the formation of a metal–acyl complex (**5.1.4 II**) by insertion of CO into the metal–alkyl complex (**5.1.4 I → 5.1.4 II**), as illustrated in Fig. 5.1.4 for the hydroformylation of an olefin by a Rh catalyst.

The metal–alkyl complex 5.4.II results from the insertion of an olefin into the Rh–H bond of the olefin complex (Fig. 5.1.4 IV). This latter insertion step is the same one which occurs in olefin polymerization on Ziegler–Natta and Metallocene catalysts (Sec. 4.1.9.) In the hydroformylation case (5.1.4) the metal–acyl complex (**5.1.4 II**) reacts with H_2 in a reaction analogous to a polymerization termination by hydrogen which forms a saturated end group (as in 4.1.9).

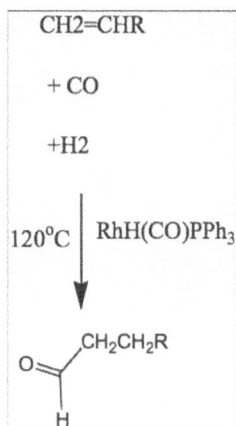

Ref: Moulijn

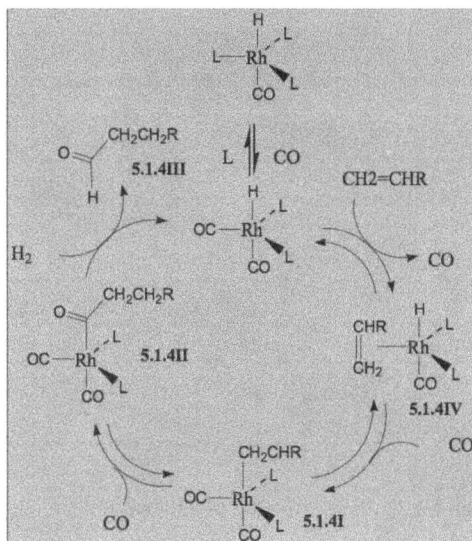

Fig. 5.1.4 Hydroformylation: CO Insertion (Ref. 1, reproduced with permission)

5.1.5 *Synthesis Gas Reactions*

The mixture of CO and H_2, known as *synthesis gas,* is an important raw material for synthesis of a number of industrially important chemicals and fuels, including methanol, methane, higher hydrocarbons and oxygenates. Synthesis gas can be produced by gasification of coal, and therefore provides a non-petroleum based feedstock for production of fuels and chemicals. For this reason, the fuels produced from synthesis gas are often referred to as "synthetic fuels" or "synfuels."

Synthesis gas can also be formed from petroleum-based liquid hydrocarbons and methane (currently the major source of synthesis gas) by steam reforming or partial oxidation. The CO portion of synthesis gas can be further converted to H_2 by use of the water–gas shift reaction (Fig. 5.1.2.2). The key reactions of involving synthesis gas are summarized in Fig. 5.1.5:

(1) *Steam Reforming* of methane: conversion of CH_4 and H_2O to CO/H_2,
(2) *Methanation*: conversion of CO/H_2 to CH_4,

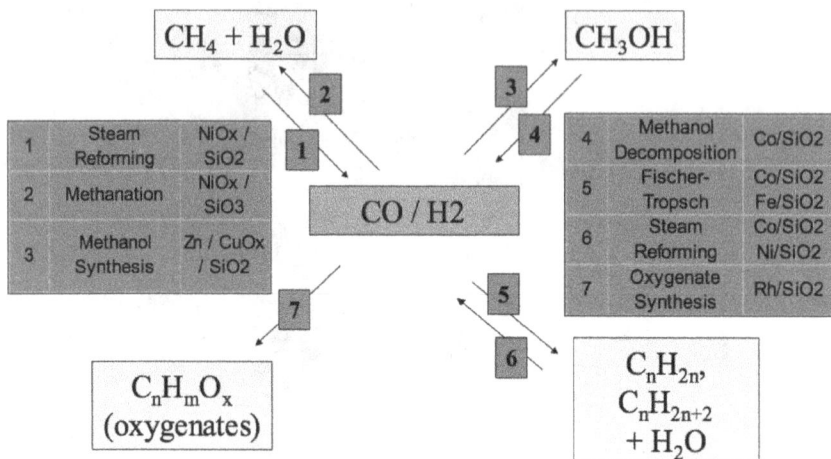

Fig. 5.1.5 **Synthesis Gas Reactions** (Refs. 1, 40 reproduced with permisssion)

(3) *Methanol Synthesis*: conversion of CO/H_2 to methanol,

(4) *Methanol Decomposition*: conversion of CH_3OH to CO/H_2

(5) *Fischer–Tropsch* Reaction: conversion of CO/H_2 to gasoline–range hydrocarbons,

(6) *Steam Reforming* of hydrocarbons: conversion of hydrocarbons and steam to CO/H_2,

(7) *Oxygenate Synthesis*: conversion of CO/H_2 to higher oxygenated hydrocarbons.

NiO/SiO_2 catalysts are used for steam reforming and methanation, $Zn/Co/SiO_2$ for methanol synthesis, Co/SiO_2 for methanol decomposition, Fischer–Tropsch and steam reforming, Fe/SiO_2 for Fischer–Tropsch, and Rh/SiO_2 for oxygenate synthesis. These reactions will be discussed in more detail subsequently in this Chapter.

5.1.6 Synthesis Gas: Non-Dissociative CO Adsorption: MeOH

One of the main determinants of which product will be formed from synthesis gas is whether or not the catalyst will dissociate CO. Non-dissociative CO adsorption is believed to result in methanol formation,

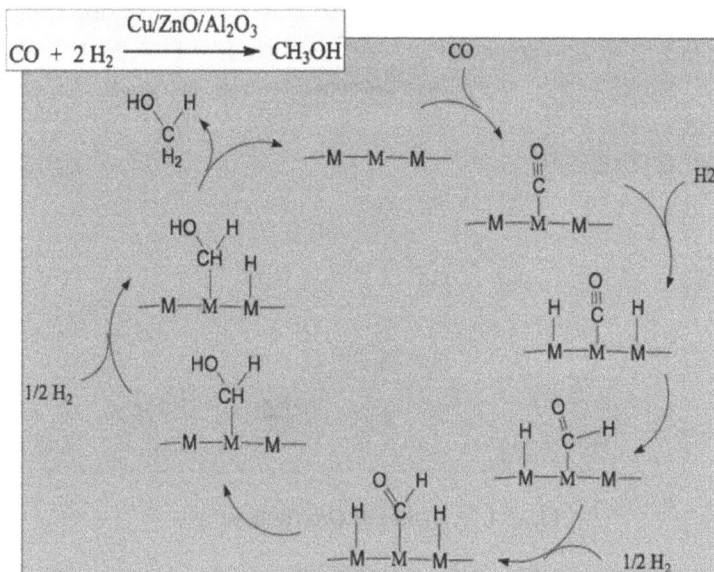

Fig. 5.1.6   Synthesis Gas (CO/H$_2$): Non-dissociative CO Adsorption: MeOH (Refs. 40, 41)

while dissociative CO adsorption produces hydrocarbons via the Fischer–Tropsch reaction.

The non-dissociative adsorption of CO results in methanol formation, as depicted in Fig. 5.1.6. This occurs on Cu/Zn/SiO$_2$ catalyst. Adsorbed CO is consecutively hydrogenated to form acyl (–CHO), hydroxymethyl (CH$_2$OH) and finally desorbed methanol. CO is not dissociated into M–C and M–O species, and this is therefore considered a non-dissociative process.

5.1.7 *Synthesis Gas: Dissociative CO Adsorption: Fischer–Tropsch Hydrocarbons*

The reaction of CO/H$_2$ over Fe/SiO$_2$ or Co/SiO$_2$ catalysts however, does result in the dissociation of CO on the metal surface into M–C and M–O species. More CO is dissociated on the M–C site and is subsequently hydrogenated to form a growing alkyl group bonded to the metal. Analogous to coordination olefin polymerization (4.1.9), the metal alkyl

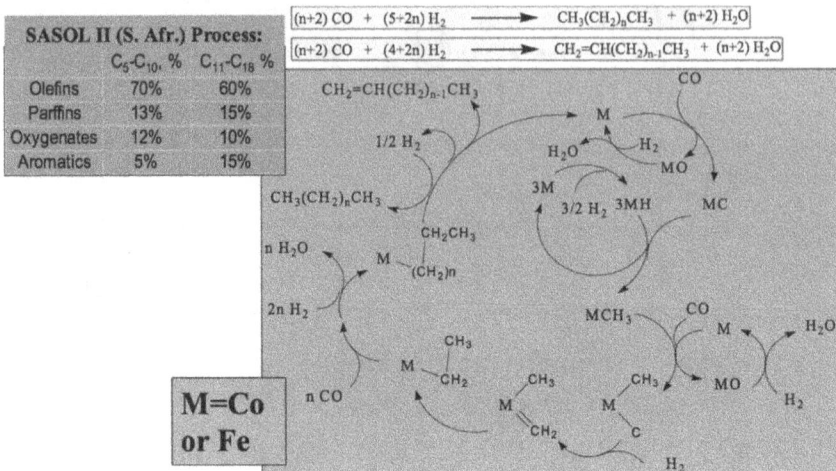

Fig. 5.1.7 Synthesis Gas (Refs. 40, 41)

can either undergo chain transfer to form an unsaturated end group or termination by H_2 to form a saturated end group. Hydrocarbons in the gasoline range (C_5–C_{18}) are formed in this process with more olefins formed than paraffins, indicative of more chain transfer than termination. Oxygenates and aromatics are also formed, which shows that some non-dissociative CO adsorption and dehydrogenation reactions are also occurring.

5.1.8 Fischer–Tropsch: Schultz–Flory Distribution

The distribution of hydrocarbons formed from Fischer–Tropsch catalysts can be explained based on the polymerization of methylene (CH_2) units with a characteristic propagation ratio called α, defined as

$$\alpha = k_p/(k_p + k_t), \tag{5.1.8.1}$$

where

k_p is the propagation rate constant and

k_t is the termination rate constant in the Schultz–Flory distribution, depicted in the upper left scheme of Eq. 5.1.8. At steady state,

$$\alpha = [C_n + 1]/[C_n], \tag{5.1.8.2}$$

where

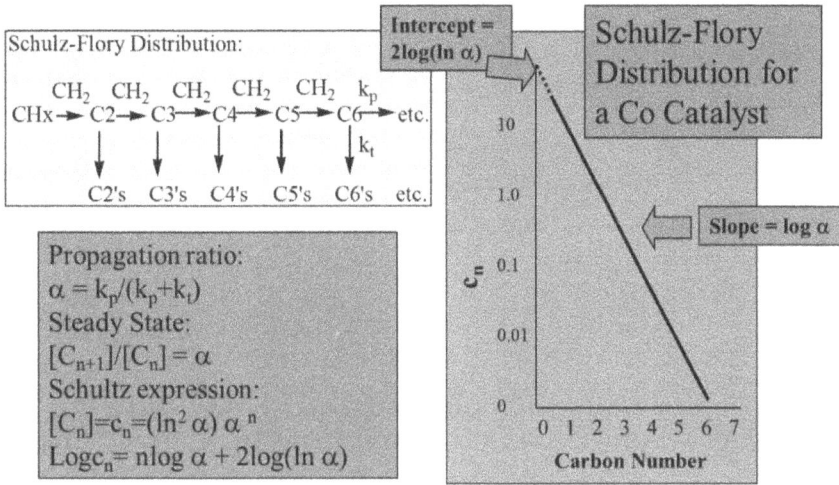

Fig. **5.1.8** **Fischer–Tropsch Distribution** (Refs. 3, 41)

C_n is the concentration of a hydrocarbon of carbon number $= n$. Under these conditions, the distribution follows the Schultz expression:

$$[C_n] = c_n = (\ln 2\ \alpha)\ \alpha^n \text{ or} \tag{5.1.8.3}$$

$$\log c_n = n \log \alpha + 2 \log (\ln \alpha), \tag{5.1.8.4}$$

which defines a straight line in a plot of c_n versus carbon number (n) with slope — $\log \alpha$ and y-intercept $= 2 \log (\ln \alpha)$. Each catalyst and set of conditions will have a characteristic α, which, for a Co catalyst, is defined by the distribution shown in the plot on the side of right of Eq. 5.1.8.

5.1.9 *Hydrodesulfurization*

Hydrodesulfurization (HDS) is the major process by which S is removed from fuels. This is becoming increasingly important with the greater use of exhaust gas catalysts to reduce pollution from internal combustion engines, especially noble metal particulate and NO*x* reduction catalysts. These catalysts are poisoned by even minute amounts of S, hence the increasing need for ultra-low levels (<5 ppm) of S in fuels.

The predominant commercial HDS catalyst is a sulfided mixed oxide of Co and Mo supported on Al_2O_3 known as $Co/Mo/S/Al_2O_3$. This catalyst

HDS Catalyst: Co/Mo/S/Al$_2$O$_3$

$$\text{Co(NO}_3)_2 + \text{(NH}_4^+)_2\text{(MoO}_4)= \xrightarrow[\text{Al}_2\text{O}_3]{\text{H}_2\text{O}} \xrightarrow[\text{calcine}]{\text{dry}} \text{CoO /MoO}_3\text{/Al}_2\text{O}_3$$

$$\text{CoO /MoO}_3\text{/Al}_2\text{O}_3 \xrightarrow{\text{H}_2\text{S / H}_2} \text{Co}_9\text{S}_8\text{/MoS}_2\text{/Al}_2\text{O}_3 = \text{Co/Mo/S/Al}_2\text{O}_3$$

HDS Reactions:

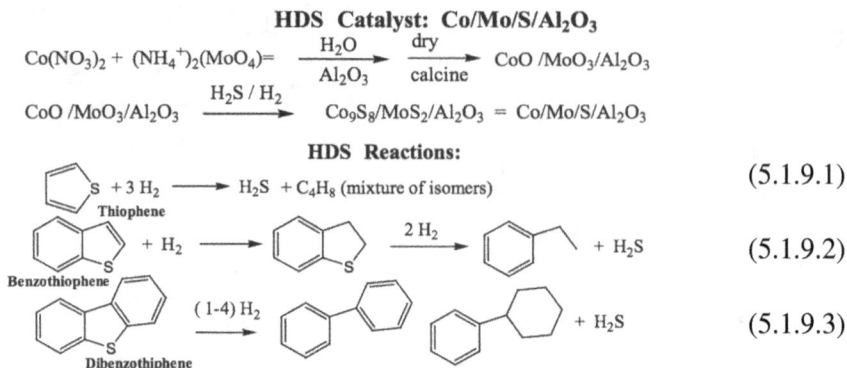

(5.1.9.1)

(5.1.9.2)

(5.1.9.3)

Fig. 5.1.9   HDS (Ref. 5)

is produced by first impregnation of Al$_2$O$_3$ with an aqueous mixture of cobalt nitrate and ammonium molybdate. After drying and calcination in air, a mixture of the corresponding oxides on the alumina support is formed. This supported mixed oxide is then sulfided by heating with a mixture of H$_2$S and H$_2$ to convert the oxides to the corresponding sulfides shown in fig. 5.1.9. Sulfur is present in hydrocarbon fuels as thiophene, benzothiophene and dibenzothiophene. The Co/Mo/S/Al$_2$O$_3$ catalyst, in the presence of hydrogen, extracts S from these compounds as H$_2$S by the reactions shown in Eqs. 5.1.9.1–5.1.9.3.

5.2 Elements of Catalyst Design

As with other types of catalysts, metals and ligand substitution and the addition of promoters and poisons are an important means of designing a reduction catalyst for a desired reaction. Chiral ligands can be used to impart stereo selectivity to catalytic hydrogenation.

5.2.1 *Metal Substitution Effect: Fe Versus Co in FT*

The effect of metals is illustrated by the comparison of the hydrocarbon distribution produced from Fe-based versus Co-based *Fischer–Tropsch* catalysts. From Eq. 5.1.8.4, the slope (m) of the C$_n$ versus carbon number

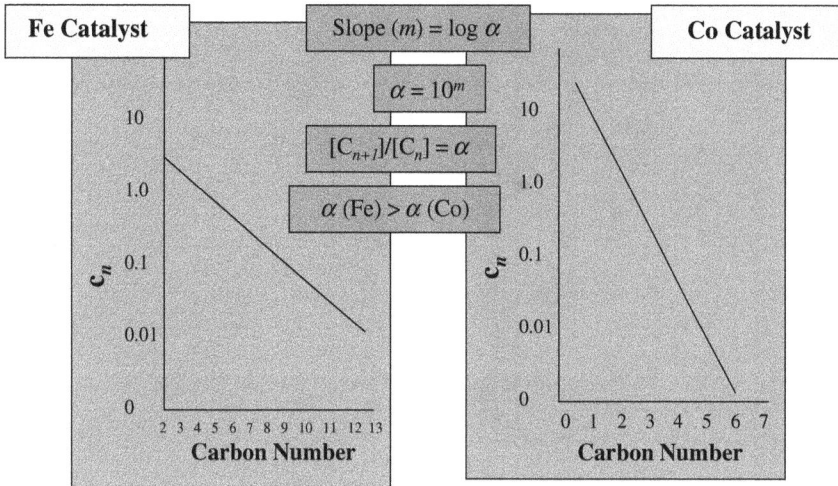

Fig. 5.2.1.1 Metal Substitution Effect: Fe versus Co in FT (Ref. 41)

(n) graph is the log of α ($\alpha = 10^m$), where α is the fraction of the total rate that is due to propagation, i.e., the rate of propagation: rate of (propagation + termination) (Eq. 5.1.8.1) or (from Eq. 5.1.8.2) the ratio of the number of molecules with $n + 1$ carbon atoms to the number of molecules with n carbon atoms (C_{n+1}/C_n).

The slope of the C_n versus carbon number (n) for Fe is greater (lower negative slope) than that for the Co catalyst. Therefore, the ∂ value for Fe is greater than that for Co, and thus, the Fe catalyst produces a larger fraction of higher molecular weight hydrocarbons than Co.

In terms of mechanism, the propagation:termination ratio is controlled by the corresponding reactions of the key metal alkyl intermediate (Fig. 5.2.1.1). Propagation (k_p) involved the addition of another CH_2 unit from reaction with Co and H_2, while termination (k_t) involved the M–C bond cleavage by reductive elimination (to produce paraffins) or chain transfer (to produce olefins). A greater fraction of the Fe catalysts follow the propagation route than do the Co catalysts, and thus, the former produces a greater fraction of higher molecular weight hydrocarbons.

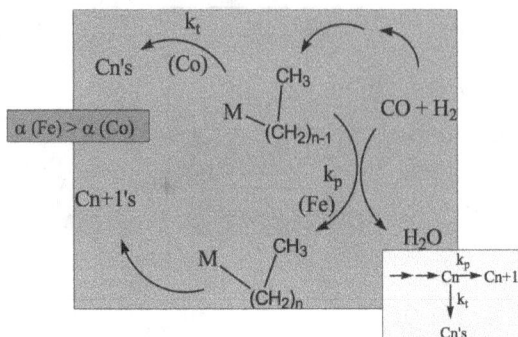

Fig. 5.2.1.1   Metal Substitution Effect; Fe Versus Co in FT (Refs. 40, 41)

5.2.2 *Metals/Ligand Effect: Hydroformylation*

The effects of metals and ligands on hydrogenation catalysts is illustrated by the fraction of linear and branched products from hydroformylation catalysts. Hydroformylation of olefins, i.e., the addition of the elements of "H" and "CHO" across an olefin double bond using synthesis gas (CO/H_2), can be catalyzed by Co and Rh catalysts. Co catalysts produce mainly the linear product (60–70%). The introduction of phosphine ligands increases the amount of linear product to 70–90%. By substitution of Rh

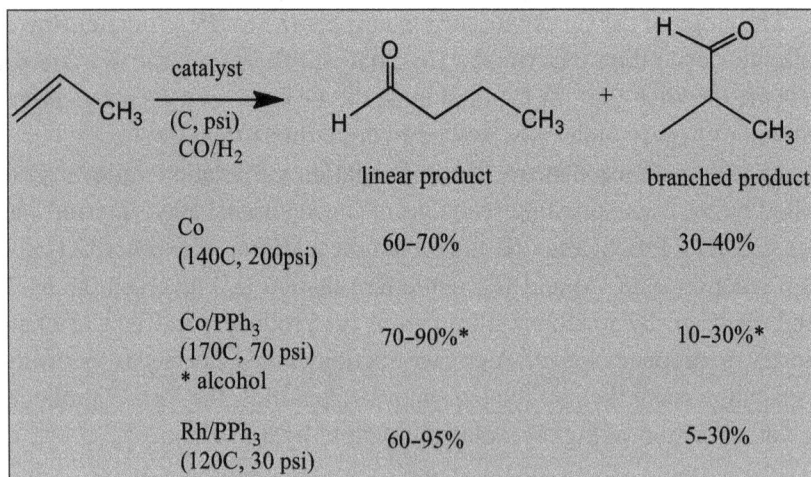

catalyst (C, psi) CO/H_2	linear product	branched product
Co (140C, 200psi)	60–70%	30–40%
Co/PPh$_3$ (170C, 70 psi) * alcohol	70–90%*	10–30%*
Rh/PPh$_3$ (120C, 30 psi)	60–95%	5–30%

Fig. 5.2.2   Metals/Ligand Effect: Hydroformylation (Ref. 1)

for Co with phosphine ligands, as high as 95% selectivity to linear product is possible.

5.2.3 Hydroformylation Ligand Effects

5.2.3.1 χ and Φ values

The ligand effect described in 5.2.2 is largely an electronic one, where the ligand donates to, or withdraws electron density from, the metal center. This electronic effect of a phosphine ligand L can be defined by the parameter, which is equal to $l–x$, where x is the infrared frequency in cm^{-1} of the C = O stretch in the complex $NiL(CO)_3$ where L = $P(t–Bu)_3$ and l is the corresponding value for any other phosphine ligand PR_3. The higher the χ, the greater the electron withdrawing effect of the substituent R. Electron-withdrawing R substituents, such as halogen, have values as high as 59cm^{-1}, while carbon substitutents are in the 0–20 range, and oxygen substitutents (phosphites) are about 20–40cm^{-1}.

Ligands can also effect the steric bulk around the active metal center. This steric effect for phosphine ligands can be quantified based on the Tolman steric parameter χ, which is the cone angle between the metal and the ligand at a distance of 2.28 Å, as shown in Fig. 5.2.3.1 for

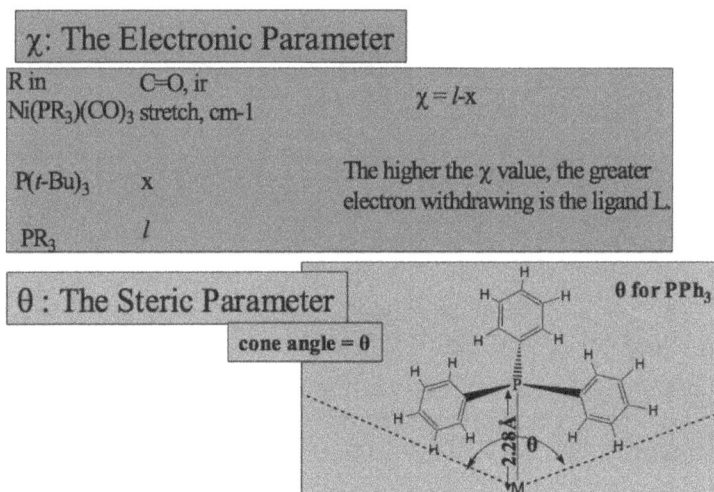

Fig. 5.2.3.1 χ and Φ **Values** (Ref. 1)

triphenyl phosphine, which has a Φ value of 145°. More bulky phosphine substituents have higher Φ values (194° as tris(2-methylphenyl) phosphine) and phosphine itself (PH_3) has a Φ value of 87°.

5.2.3.2 Electronic (χ) and steric (Φ) effects in hydroformylation

The contribution of steric and electronic effects on the linear: branched isomers of hydroformylation products are illustrated in Fig. 5.2.3.2. The hydroformylation of heptane with Rh catalysts using phosphine ligands with various χ and Φ values is shown with the corresponding distribution of linear: branched products. The linear product is generally preferred over the branched, because the linear is better for the production of oxo alcohols, which is a major use for linear aldehydes. The χ values range from 4–51 cm^{-1} and the Φ values from 109°–190°. The substituents are arranged from top to bottom in the order of decreasing % linear isomer. Higher linear isomer seems to correlate with higher χ values and, in general, mid-lower Φ values. The two cases where this does not hold are for the bottom two substituents. The [$(CF_3)_2HCO$–] substituent has such a high χ value (51), that it results

• High Linear → Low θ, High χ

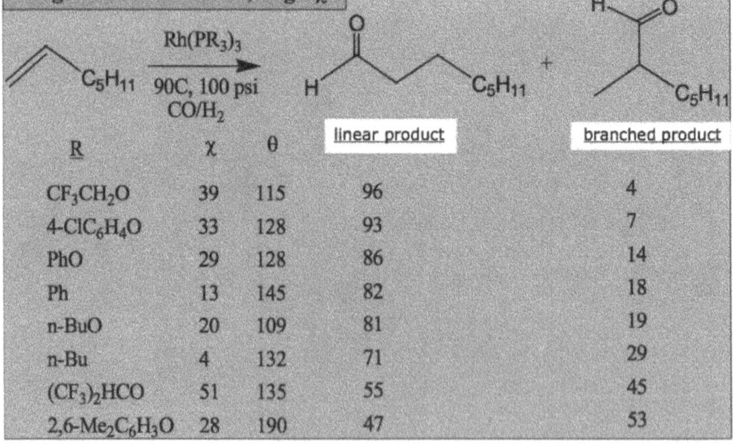

R	χ	θ	linear product	branched product
CF_3CH_2O	39	115	96	4
$4\text{-}ClC_6H_4O$	33	128	93	7
PhO	29	128	86	14
Ph	13	145	82	18
n-BuO	20	109	81	19
n-Bu	4	132	71	29
$(CF_3)_2HCO$	51	135	55	45
$2,6\text{-}Me_2C_6H_3O$	28	190	47	53

Fig. 5.2.3.2 Electronic and Steric Effects (Ref. 1)

in an (electronically) unstable Rh complex. The very bulky [2,6-Me$_2$C$_4$H$_3$O–] ligand (Φ = 190°) also destabilizes the Rh complex for steric reasons. Thus, within the stability limits of the Rh catalyst, the linear product is favored by electron-withdrawing substituents and with lower-to-medium substituent bulk.

5.2.4 *Asymmetric Hydrogenation*

Ligands that are chiral can be used to preferentially produce one stereoisomer (*stereoselectivity.*) L–DOPA, a drug used in the treatment of Parkinson's disease, is produced by a catalytic hydrogenation process which uses the chiral ligand shown in 5.2.4 II. The most favored configuration of the starting olefin 5.2.4 I with respect to the metal-ligand complex 5.2.4 II is that shown in 5.2.4 III, where the two bulky groups Ar and OAn are furthest apart (versus 5.2.4 IV which is not favored). Oxidative insertion of H$_2$ into 5.2.4. III and insertion of olefin into one of the resulting Rh–H bonds results in 5.2.4 VI. Reductive elimination produces L–DOPA (5.2.4 VII) with the stereochemistry shown.

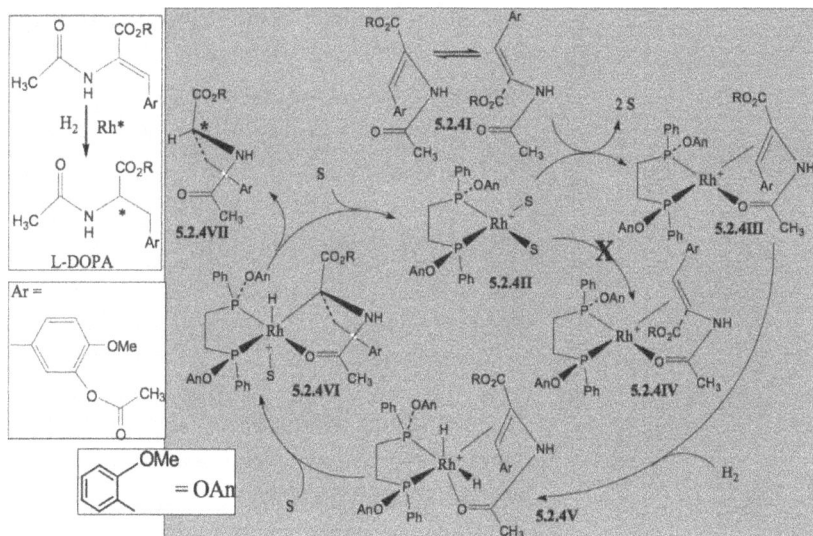

Fig. 5.2.4   Asymmetric Hydrogenation (Ref. 1)

5.2.5 **Promoters and**
Poisons: Noble Metals

• Common Poisons (strongly
  binds to M):
  — S (e.g., RSH, H₂S, R₂S)
  — CO
  — P
  — C/Coke

• Common Promoters
  (inhibits coke):
  — H₂
  — H₂O

5.2.5 *Promoters and Poisons: Noble Metals*

The noble metals, such as Pt, and Pd, that are used in reduction and hydrogenation catalysts can be poisoned by many molecules. Among these are organo–sulfur compounds or H_2S, P and, as for other catalysts, coke. CO can be both a reactive ligand and a poison, depending on the metal and the concentration of CO. Common promoters for noble metals include H_2 and H_2O, which can react with impurities and surface oxygen to maintain a reduced, clean metal surface.

5.3 Major Industrial Processes

Examples of major industrial processes that utilize reduction/hydrogenation catalysts include the refinery processes such as hydro-cracking/hydrotreating and catalytic reforming, and those for chemical production, for example dehydrogenation of ethylbenzene to styrene, hydroformylation as a source of oxo-alcohols, carbonylation of methanol to acetic acid, and ammonia synthesis by the Haber–Bosch process. These are discussed in more detail in the remainder of this chapter.

5.3.1 *Hydrogenolysis: Hydrocracking and Hydrotreating*

An important example of the commercial application of *hydrogenolysis* (hydrogenation with C–C bond breaking) is the process known as hydrocracking/hydrotreating, by which hydrocarbons are converted to smaller paraffins and olefins. This is very valuable for converting

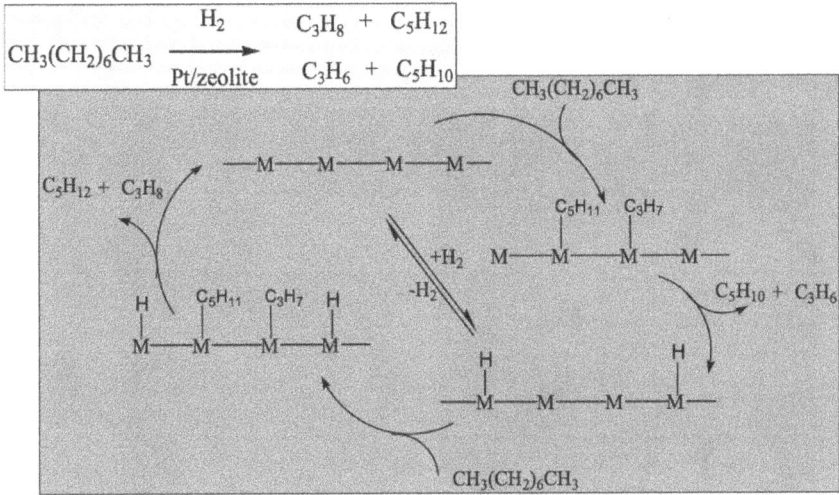

Fig. 5.3.1 Hydrogenolysis: Hydrocracking/Hydrotreating (Refs. 1, 18)

less useful heavy hydrocarbons into lighter ones that can be used in liquid fuels such as gasoline. A representative reaction is the conversion of n–octane to a mixture of C_3 and C_5 olefins and paraffins. A reduced noble metal surface such as Pt serves as a site for reductive cleavage of octane to C_5 and C_3 metal alkyls.

β-hydride elimination produces the corresponding C_5 and C_3 olefins and a metal-hydride surface. Subsequent dissociative chemisorption of octane and reductive elimination of the resulting C_5 and C_3 metal alkyl/ metal hydride surface species produces the corresponding C_5 and C_3 paraffins. In general, the overall process is important in refinery operations by production of lighter, fuel grade olefins and paraffins from lower value, heavier hydrocarbons.

5.3.2 *Dehydrogenation: Styrene*

Hydrogenation catalysts are also catalysts for dehydrogenation for production of stable olefins such as styrene, where the loss of hydrogen from the corresponding paraffin (in this case, ethylbenzene) is thermodynamically possible. The reaction is reversible, and thus high temperatures (600°C) drive the equilibrium of Eq. 5.3.2 to the right, which has a positive

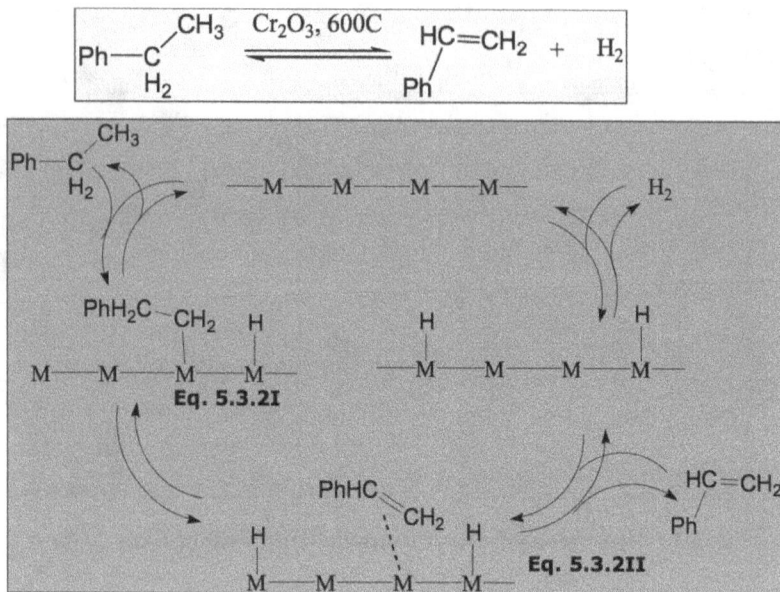

Fig. 5.3.2 **Dehydrogenation** (Refs. 1, 18)

ΔS because two molecules (H_2 and styrene) are formed from one (ethylbenzene). When styrene is allowed to pass over a CrO_3 catalyst at 600°C, an equilibrium with ethybenzene and H_2 occurs *via* the reversible reductive chemisorption of ethylbenzene to the corresponding metal-alkyl/metal hydride species 5.3.2 I and subsequent β-elimination to the styrene-dihydride surface 5.3.2 II. The latter is in equilibrium with styrene and hydrogen. Again, all these reactions are reversible and are driven to styrene + hydrogen by high temperatures.

5.3.3 *Hydroformylation/Hydrogenation: Oxo Alcohols*

The most commercially significant use of the aldehydes formed by hydroformylation, either by Rh (Fig. 5.1.4) or Co catalysts (Fig. 1.6.5) is for the production of oxo-alcohols by *aldol condensation*, which is the reaction of two molecules of enolizable aldehyde molecules with loss of water to the corresponding a,b-unsaturated ketone. The mechanism for the base-catalyzed aldol condensation is shown in Fig. 5.3.3. The most prominent reaction in this class is for production of the C_8 a,b-unsaturated ketone in Fig. 5.3.3I from two molecules of butyraldehyde. Recall that

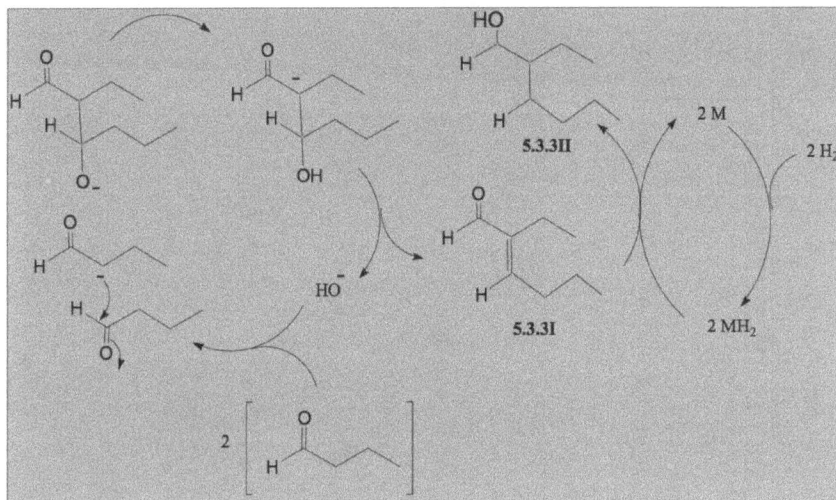

Fig. 5.3.3 Butanal Aldol Condensation To 2-Ethylhexanol (Refs. 1, 18)

butyraldehyde is produced as the linear product from the hydroformylation of propylene (5.2.2).

The desired oxo-alcohol 2-ethylhexanol (5.3.3II) is produced by catalytic hydrogenation of the aldol product 5.3.1I over a metal catalyst (M) by the mechanism shown in Fig. 5.3.3.

5.3.4 *Carbonylation: Methanol to Acetic Acid*

Carbonylation of methanol is another means of using CO as a building block for *homologazation* (increasing carbon number by 1). CO can be inserted into the C–O bond of CH_3OH to produce acetic acid, using a CoI_2 catalyst. Mechanistically, the reaction is similar to hydroformylation in that a metal acyl complex 5.3.4 I is formed as an intermediate (Fig. 5.3.4.1).

In this case, instead of reaction with H_2 to form aldehyde, reaction with water forms the carboxylic acid, in this case acetic acid. The metal acyl complex results from insert of CO into the metal methyl complex 5.3.4 II. The latter is believed to result from reaction of the active catalytic species $HCo(CO)_4$ with CH_3I, which forms from CH_3OH and HI. The latter two molecules are produced. along with $HCo(CO)_4$. by reaction of CoI_2 with water and CO. The drawback to the use of Co catalysts for

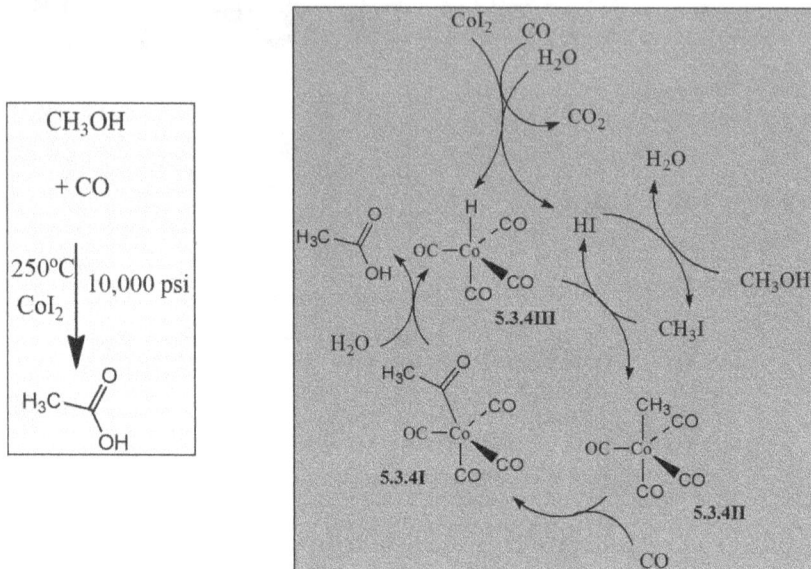

Fig. 5.3.4.1   Carbonylation: Methanol to Acetic Acid on Co Catalysts (Refs. 1, 18)

this process is the high pressure of CO (10,000 psi) that is required to produce the active $HCo(CO)_4$ catalyst.

In this case, instead of reaction with H_2 to form aldehyde, reaction with water forms the carboxylic acid, in this case acetic acid. The metal acyl complex results from insert of CO into the metal methyl complex 5.3.4 II. The latter is believed to result from reaction of the active catalytic species $HCo(CO)_4$ with CH_3I, which forms from CH_3OH and HI. The latter two molecules are produced along with $HCo(CO)_4$. by reaction of CoI_2 with water and CO. The drawback to the use of Co catalysts for this process is the high pressure of CO (10,000 psi) that is required to produce the active $HCo(CO)_4$ catalyst.

Rh forms an analogous complex, $HRh(CO)_3I_2$, at much lower pressures (500 psi), which catalyzes the carbonylation of methanol to acetic acid by a similar mechanism (shown in Fig. 5.3.4.2). For this reason, RhI_2 is a preferred to CoI_2 as a catalyst, although it is much more expensive.

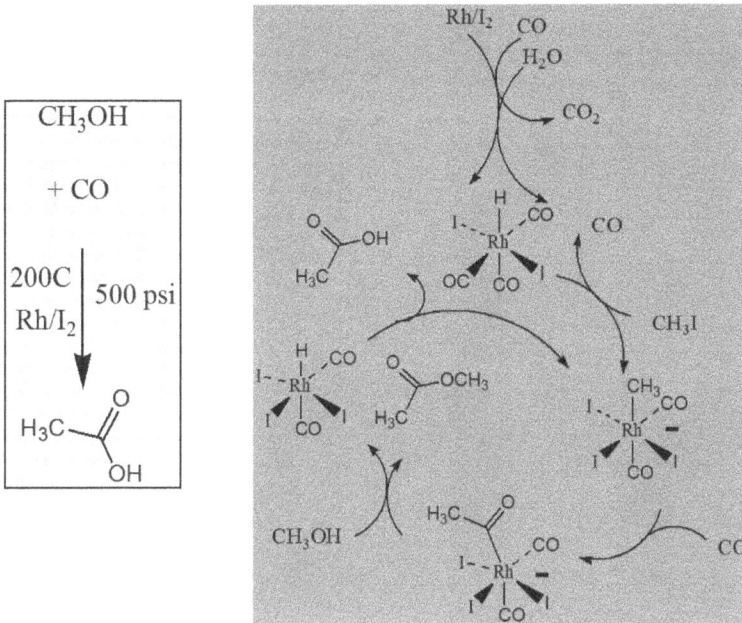

Fig. 5.3.4.2   Carbonylation: Methanol to Acetic Acid on Rh Catalysts (Refs. 1, 5, 6, 18)

5.3.5 *Catalytic Reforming*

Catalytic reforming takes place on a noble metal (Pt) catalyst supported on an acid catalyst such as a zeolite. The process combines metal-catalyzed dehydrogenation and hydrogenation reactions with acid-catalyzed isomerization reactions to produce branched paraffins, olefins and cyclic paraffins and aromatics from straight chain hydrocarbons. These products have higher octane value and are, in general, higher quality fuels than the corresponding straight-chain hydrocarbon. Thus, it is an important process for the production of gasoline in a modern refinery. A model reaction for the catalytic reforming process is the conversion of *n*-hexane to 2-methylpentane, 2-methylpentene, cyclohexane and benzene. Mechanistically, the reaction proceeds by the reaction of the *n*-hexane at high temperature (500°C) with the silica–alumina acid site of the zeolite

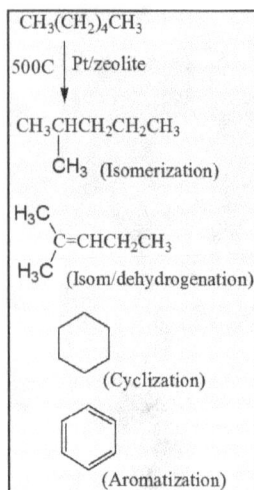

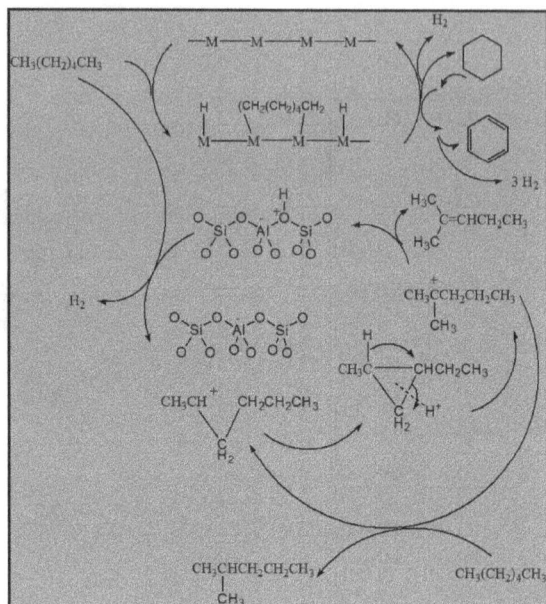

Fig. 5.3.5   Catalytic Reforming (Refs. 1, 5, 18)

to form H_2 and a 2° carbenium ion, which can rearrange and lose H+ as shown, to produce the isoolefin (2-methylpentene). Alternatively, the 2° carbenium ion can abstract a hydride from the starting *n*-hexane to form the isoparaffin (2-methylpentane) and regenerate the carbenium ion. The starting *n*-hexane can also react with the reduced metal sites to form the metal-$(CH_2)_6$ species and 2 metal hydride sites. These species eliminate H_2 and cyclohexane, which can dehydrogenate on the reduced metal sites to form H_2 and benzene.

5.3.6 *Ammonia Synthesis*

The Haber–Bosch process is one of the first large scale catalytic processes to be commercialized. It allowed high volumes of inexpensive "fixed" nitrogen to become available for use as fertilizer, explosives and dyes. The process involves the hydrogenation of N_2, by the reductive cleavage of the $N \equiv N$ triple bond to form an intermediate metal nitride, which is successively hydrogenated to ammonia, as shown in Fig. 5.3.5.

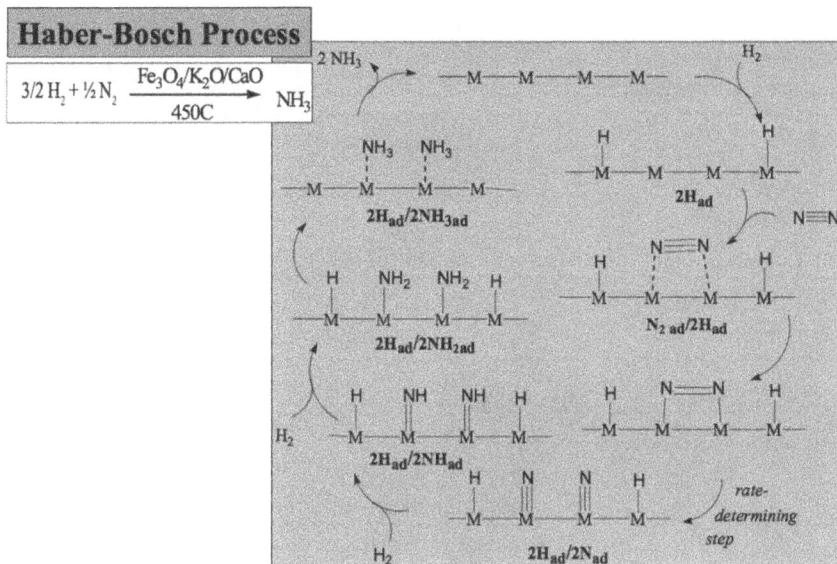

Haber-Bosch Process

$$3/2\,H_2 + \tfrac{1}{2}N_2 \xrightarrow[\;450C\;]{Fe_3O_4/K_2O/CaO} NH_3$$

Fig. 5.3.6 **Ammonia Synthesis** (Ref. 40)

The activation of H_2 by the metal surface, in this case Fe, is similar to that for olefin hydrogenation (5.1.3).

5.4 Problems

1. Match a process from Column A with a reaction from Column B and a catalyst from Column C:

A	B	C
(a) Water–Gas Shift Reaction	(1) $CO + 2H_2 \rightarrow CH_3OH$	(i) Pt/zeolite
(b) Steam Reforming	(2) $CO + 3H_2 \rightarrow CH_4 + H_2O$	(ii) Ni/SiO$_2$
(c) Methanol Synthesis	(3) $CH_3OH + CO \rightarrow CH_3CO_2H$	(iii) Cr$_2$O$_3$
(d) Fischer–Tropsch	(4) $CO + H_2O \rightarrow H_2 + CO_2$	(iv) Fe$_2$O$_3$
(e) Methanation	(5) $nCO + 2nH_2 \rightarrow CnH_{2}n + nH_2O$ $nCO + nH_2 \rightarrow CnHn + n/2H_2O$	(v) NiO$_x$/SiO$_2$
(f) Methanol Decomposition	(6) $CH_4 + H_2O \rightarrow 3H_2 + CO$	(vi) Co/SiO$_2$ or Fe/SiO$_2$

(Continued)

(*Continued*)

A	B	C
(g) Hydroformylation	(7) $CH_2(CH_2)_6CH_3 + H_2 \rightarrow C_3H_6 +$ $C_5H_{12} + C_3H_8 + C_5H_{10}$	(vii) Co/SiO_2
(h) Methanol Carbonylation	(8) $N_2 + 3/2H_2 \rightarrow NH_3$	(viii) $RhH(CO)_3I_2$
(i) Hydrocracking/ Hydrotreating	(9) $PhCH_2CH_3 \rightarrow PhCH = CH_2 + H_2$	(ix) $Fe_3O_4/K_2O/$ CaO
(k) Dehydrogenation	(10) $RCH = CH_2 + CO + H_2 \rightarrow$ RCH_2CH_2CHO	(x) $RhH(CO)$ $[PPh_3]_2$
(l) Catalytic Reforming	(11) $CH_3OH \rightarrow CO + 2H_2$	(x) $Zn/Cu/SiO_2$
(m) Ammonia Synthesis	(12) $CH_3(CH_2)_4CH_3 \rightarrow$ $(CH_3)_2CHCH_2CH_2CH_3 +$ $(CH_3)C = CHCH_2CH_3 +$ cyclohexane + benzene	

2. Match the activation step in Column B and the key intermediate–forming step in Column C with each of the Processes in Column A:

A	B	C
(a) Methanol Synthesis	(i) $CH_3RhI_3(CO)_2 + CO \rightarrow$ $CH_3(C = O)RhI_3(CO)_2$	(1) $PhCH_2$--CH_2--M--M--H $\rightarrow PhCH = CH_2 + H$-- M--M--H
(b) Fischer–Tropsch	(ii) $M + CO \rightarrow M$--$C = O$	(2) H--$M(= NH)M(= NH)$ M--$H + H_2 \rightarrow H$--$M(NH_2)$ $M(NH_2)M$--H
(c) Hydroformylation	(iii) $HRh(CO)L_2(CH_2 =$ $CHR) + CO \rightarrow$ $Rh(CO)_2L_2(CH_2CH_2R)$	(3) $CH_3(C = O)RhI_3(CO)_2 +$ $CH_3OH \rightarrow CH_3CO_2CH_3 +$ $HRhI_3(CO)_2$
(d) Methanol Carbonylation	(iv) $2M + CO \rightarrow MC + MO$	(4) $Rh(CO)_2L_2(CH_2CH_2R) +$ $CO \rightarrow Rh(CO)_2L_2$ $[C(= O)CH_2CH_2R]$
(e) Hydrocracking and Hydrotreating	(v) M--M--M--$M +$ $CH_3(CH_2)_4CH_3$-- $\rightarrow H$-- MM--$(CH_2)_6$--M--M--H	(5) C_5H_{11}--M--M--$C_3H7 \rightarrow$ H--M--M--$H + C_5H_{10} +$ C_3H_6
(f) Dehydrogenation	(vi) H--M--$N = N$--M--$H +$ $H_2 \rightarrow H$--$M(= NH)$ $M(= NH)M$--H	(6) H--M--M--$(CH_2)_6$--M-- M--$H \rightarrow$ cyclo--C_6H_{12}

(*Continued*)

(Continued)

A	B	C
(g) Catalytic Reforming	(vii) $M\text{--}M + CH_3(CH_2)_6CH_3 \rightarrow$ $\rightarrow C_5H11\text{--}M\text{--}M\text{--}C_3H_7$	(7) $H\text{--}M\text{--}M\text{--}C = O \rightarrow$ $M\text{--}M\text{--}C(=O)H$
(h) Ammonia Synthesis	(viii) $M\text{--}M + PhCH_2CH_3 \rightarrow$ $PhCH_2\text{--}CH_2\text{--}M\text{--}M\text{--}H$	(8) $RM = CH_2 \rightarrow M\text{--}CH_2\text{--}R$

3. Fill in the structures of the intermediates in olefin hydrogenation over a Pt metal catalyst:

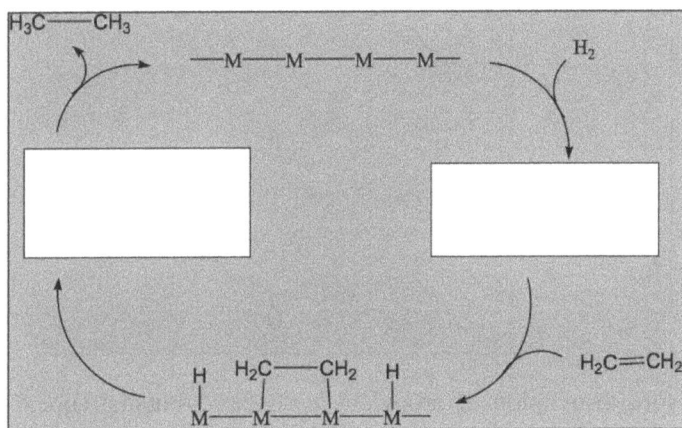

4. Using the following Schultz–Flory plots:

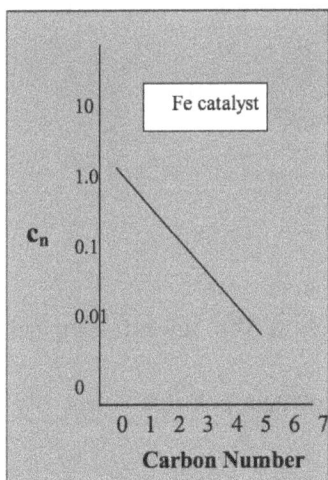

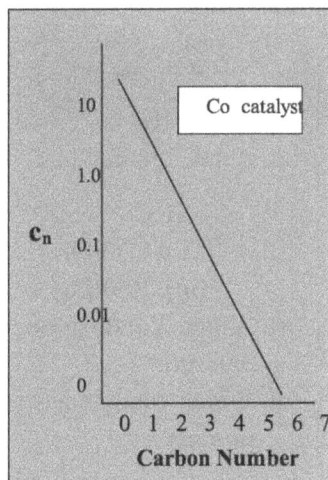

(a) Which catalyst has the higher α?

(b) What does that mean in terms of termination (kt) versus propagation (kp) rates?

   (a) $kp/(kt + kp)$ for the one with higher $\alpha > kp/(kt + kp)$ for the one with lower α,

   (b) kp for the one with higher $\alpha > kp$ for the one with lower α,

   (c) $kp/(kt + kp)$ for the one with higher $\alpha < kp/(kt + kp)$ for the one with lower α,

   (d) kp/kt for the one with higher $\alpha < kp/kt$ for the one with lower α.

(c) Identify the steps corresponding to kp and kt in the reaction below:

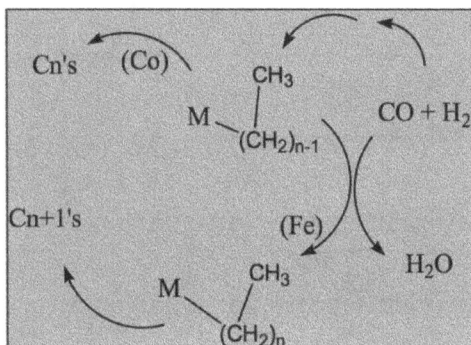

5. Hydroformylation of propylene produces 1-butanal (linear product) and 2-methyl propanal (branched product):

For the 3 catalysts below:

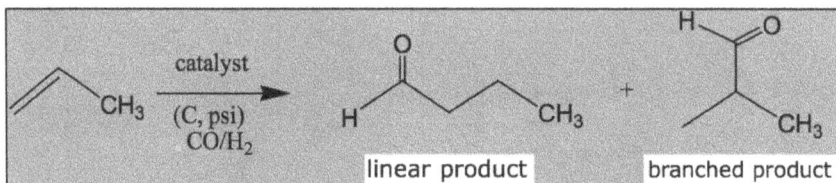

   (i) Co

   (ii) Co/PPh$_3$

   (iii) Rh/PPh$_3$

(a) Which catalyst operates at the lowest temperature and pressure?

(b) Which catalyst has the greatest possible selectivity to linear product?

For the 3 catalysts of the formula: $Rh(PR_3)_3$, where

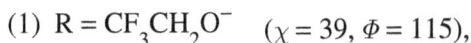

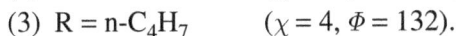

(1) $R = CF_3CH_2O^-$ $\quad(\chi = 39, \Phi = 115)$,

(2) $R = Ph$ $\quad\quad\quad\;(\chi = 13, \Phi = 145)$,

(3) $R = n\text{-}C_4H_7$ $\quad\;(\chi = 4, \Phi = 132)$.

    (c) Which catalyst has the greatest electron withdrawing ligand?

    (d) Which catalyst has the greatest steric bulk?

    (e) Which catalyst produces the greatest amount of linear product?

6. For the model reforming reaction:

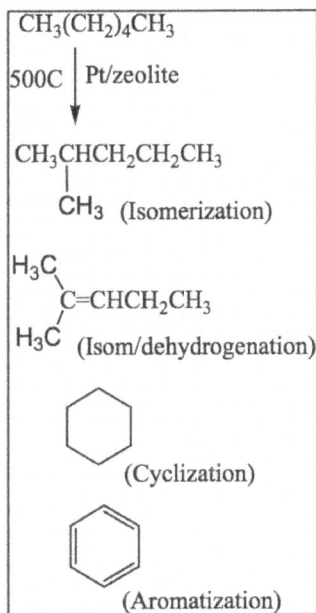

    (a) Which of the catalytic elements is responsible for the isomerization reactions?

    (b) Which of the catalytic elements is responsible for the cyclization and aromatization reactions?

    (c) Why is this an important refinery process in the production of gasoline?

5.5 Answers to Problems

1.

A	B	C
(a) Water Gas Shift	(4) $CO + H_2O \rightarrow H_2 + CO_2$	(iv) Fe_2O_3
(b) Steam Reforming	(6) $CH_4 + H_2O \rightarrow 3H_2 + CO$	(ii) Ni/SiO_2
(c) Methanol Synthesis	(1) $CO + 2H_2 \rightarrow CH_3OH$	(xi) $Zn/CuO_x/SiO_{x2}$

(Continued)

(*Continued*)

A	B	C
(d) FIscher–Tropsch	(5) $nCO + 2nH_2 \rightarrow CnH_2n + nH_2O$ $nCO + nH_2 \rightarrow CnHn + n/2H_2O$	(vi) Co/SiO$_2$, Fe/SiO$_2$
(e) Methanation	(2) $CO + 3H_2 \rightarrow CH_4 + H_2O$	(v) NiOx/SiO$_2$
(f) Methanol Decomposition	(11) $CH_3OH \rightarrow CO + 2H_2$	(vii) Co/SiO$_2$
(g) Hydroformylation	(10) $RCH = CH_2 + CO + H_2 \rightarrow RCH_2CH_2CHO$	(x) RhH(CO)[PPh3]$_2$
(h) Methanol Carbonylation	(3) $CH_3OH + CO \rightarrow CH_3CO_2H$	(viii) RhH(CO)$_3$I$_2$
(i) Hydrocracking and Hydrotreating	(7) $CH_3(CH_2)6CH_3 + H_2 \rightarrow C_3H_6 + C_5H_{12} + C_3H_8 + C_5H_{10}$	(i) Pt/zeolite
(k) Dehydrogenation	(9) $PhCH_2CH_3 \rightarrow PhCH = CH_2 + H_2$	(iii) Cr$_2$O$_3$
(l) Catalytic Reforming	(12) $CH_3(CH_2)4CH_3 \rightarrow (CH_3)2CHCH_2CH_2CH_3 + (CH_3)C = CHCH_2CH_3 +$ cyclohexane + benzene	(i) Pt/zeolite
(m) Ammonia Synthesis	(8) $N_2 + 3/2H_2 \rightarrow NH_3$	(ix) Fe$_3$O$_4$/K$_2$O/CaO

2.

A	B	C
(a) Methanol Synthesis	(ii) $M + CO \rightarrow M–C \equiv O$	(7) $H–M–M–C \equiv O \rightarrow M–M–C(= O)H$
(b) Fischer–Tropsch	(iv) $2M + CO \rightarrow MC + MO$	(8) $RM = CH_2 \rightarrow M–CH_2–R$
(c) Hydroformylation	(iii) $HRh(CO)L_2$ $(CH_2 = CHR) + CO \rightarrow Rh(CO)_2L_2 (CH_2CH_2R)$	(4) $Rh(CO)_2L_2(CH_2CH_2R) + CO \rightarrow Rh(CO)_2L_2 [C(= O)CH_2CH_2R]$
(d) Methanol Carbonylation	(i) $CH_3RhI_3(CO)_2 + CO \rightarrow CH_3(C = O) RhI_3(CO)_2$	(3) $CH_3(C = O)RhI_3 (CO)_2 + CH_3OH \rightarrow CH_3CO_2CH_3 + HRhI_3(CO)_2$

(*Continued*)

(*Continued*)

A	B	C
(e) Hydrocracking and Hydrotreating	(vii) $M–M + CH_3(CH_2)6CH_3$ $\rightarrow C_5H_{11}–M–M–C_3H_7$	(5) $C_5H_{11}–M–M–C_3H_7$ $\rightarrow H–M–M–H + C_5H_{10}$ $+ C_3H_6$
(f) Dehydrogenation	(viii) $M–M + PhCH_2CH_3$ $\rightarrow PhCH_2–CH_2–M–$ $M–H$	(1) $PhCH_2–CH_2–M–M–H$ $\rightarrow PhCH = CH_2 +$ $H–M–M—H$
(g) Catalytic Reforming	(v) $M–M–M–M +$ $CH_3(CH_2)_4CH_3$ $\rightarrow H–M–M–(CH_2)_6–$ $M–M–H$	(6) $H–M–M–(CH_2)_6–M–$ $M–H \rightarrow cyclo–C_6H_{12}$
(h) Ammonia Synthesis	(vi) $H–M–N = N–M–H +$ $H_2 \rightarrow H–M(= NH)$ $M(= NH)M–H$	(2) $H–M(= NH)M(= NH)$ $M–H + H_2 \rightarrow$ $H–M(NH_2)M(NH_2)$ $M–H$

3.

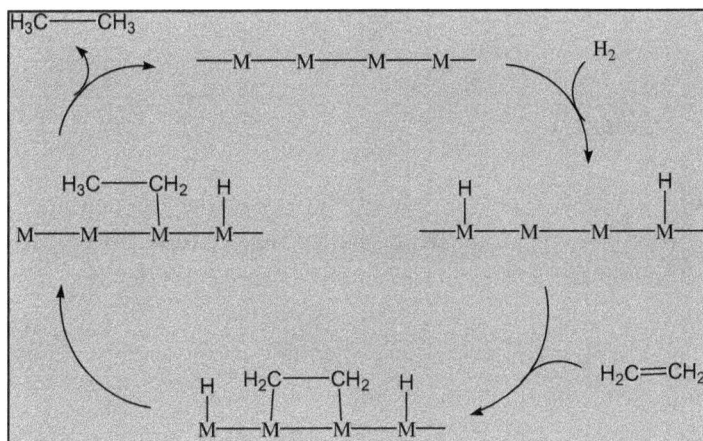

4.

   (a) Fe $kp/(kt + kp)$ for the one with higher $\alpha > kp/(kt + kp)$ for the one with lower α.

(b) Identify the steps corresponding to *kp* and *kt* in the reaction below:

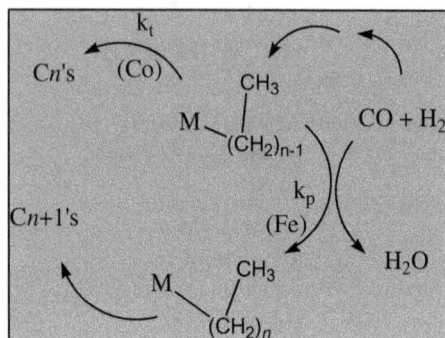

5.

    (a) iii,

    (b) iii,

    (c) 1. $R = CF_3CH_2O-$ ($\chi = 39$),

    (d) 2. $R = Ph$ ($\theta = 145$),

    (e) 1. $R = CF_3CH_2O-$ ($\chi = 39$, $\theta = 115$).

6.

    (a) Zeolite,

    (b) Pt,

    (c) This is an important refinery process in the production of gasoline because it produces higher octane hydrocarbon components versus the less environmentally friendly oxygenates.

6

Environmental Catalysis

The reduction of environmental pollutants is one of the greatest accomplishments of catalytic chemistry. Recent legislation in North America, Europe and Japan has required dramatic reduction of pollutants in exhaust gas, particularly particulates (mainly soot), NO_x, hydrocarbons and CO. This applies to both mobile and stationary applications, but the toughest problems are in the mobile area, which will be emphasized in this Chapter. For example, the simultaneous reduction of soot and NO_x in diesel fuel posed an especially difficult technical problem because of the levels of reduction that are being mandated at about 90% compared to historic levels. Nearly all of the practical technologies to effect these reductions involve catalysts. This provides an important opportunity area for the chemist interested in a career in industrial catalysis.

This chapter will discuss the elements involved in environmental catalysts, thermodynamic limitations, the various pollutant components, their reaction chemistry, and the corresponding environmental catalysts and their mechanisms of operation.

As for the other catalyst systems, the design of environmental catalysts is based on the choice of metals, the synthetic method, and their engineering into a practical system. The requirement for catalysts with a long lifetime (450,000 miles for certain diesel exhaust catalysts) means that particular attention must be paid to the effect of poisons and the use of promoters to improve catalyst activity and life.

In place of the catalytic processes to produce high volume chemicals and fuels, in this section is discussed the major types of environmental catalysts and the mechanisms employed to reduce the various exhaust gas pollutants, along with a brief summary of catalysts for fuel cells.

6.1 Concepts

Catalysts for exhaust gas conversion (Table 6.1.1) fall into four categories:

Noble metal catalysts, such as Pt, Pd and Rh
Base metal catalysts, such as V, Fe, Co, Ni and Cu
Alkaline earth metals, such as Mg, Ca, Sr and Ba
Rare earth metals, such as La and Ce

Alumina is the main catalyst support for Nobel metals and silica for the base metals. The functions of each of these catalytic elements will be discussed in detail.

Table 6.1.1:   Exhaust Gas Catalyst Periodic Table

1a	2a	3b	4b	5b	6b	7b		8		1b	2b	3a	4a	5a	6a	7a	0
H	He																
Li	Be											B	C	N	O	F	Ne
Na	Mg											Al	Si	P	S	Cl	Ar
K	Ca	Sc	Ti	V	Cr	Mn	Fe	Co	Ni	Cu	Zn	Ga	Ge	As	Se	Br	Kr
Rb	Sr	Y	Zr	Nb	Mo	Te	Ru	Rh	Pd	Ag	Cd	In	Sn	Sb	Te	I	Xe
Cs	Ba	La*	Hf	Ta	W	Re	Os	Ir	Pt	Au	Hg	Tl	Pb	Bi	Po	At	Rn
Fr	Ra	Ac**															
*Lanthanides	Ce	Pr	Nd	Pm	Sm	Eu	Gd	Tb	Dy	Ho	Er	Tm	Y	Lu			
**Actinides	Th	Pa	U	Np	Pu	Am	Cm	Bk	Cf	Es	Fm	Md	No	Lw			

6.1.2. Thermodynamic Considerations

One of the major thermodynamic limitations is due to the preferential oxidation of hydrocarbons or CO with O_2 versus NO (Fig. 6.1.2). That is the $-\Delta H$ for oxidation of CO with O_2 ($-\Delta H_{COox}$) is greater than that for the corresponding oxidation with NO ($-\Delta H_{NOredCO}$). Likewise, the $-\Delta H$

$$CO + \tfrac{1}{2}O_2 \rightarrow CO_2 \qquad\qquad\qquad \Delta H_{COox}$$
$$C_nH_m + (n+m/4)\,O_2 \rightarrow nCO_2 + m/2H_2O \qquad \Delta H_{HCox}$$
$$CO + NO \rightarrow \tfrac{1}{2}N_2 + CO_2 \qquad\qquad\qquad \Delta H_{NredCO}$$
$$C_nH_m + 2n+m/2NO \rightarrow n+m/4N_2 + nCO_2 + m/2H_2O\ \Delta H_{NredHC}$$

$$-\Delta H_{COox} > -\Delta H_{NredCO} \qquad\qquad -\Delta H_{HCox} > -\Delta H_{NredCO}$$

Gasoline (spark ignition) Engines:
→ O_2 level controlled by sensors
→ Simultaneous removal of C_nH_m, CO and NO.

Diesel (compression ignition) Engines:
→ Excess O_2 present in exhaust gas.
→ Removal of C_nH_m, CO and NO requires cycling of lean/rich conditions.

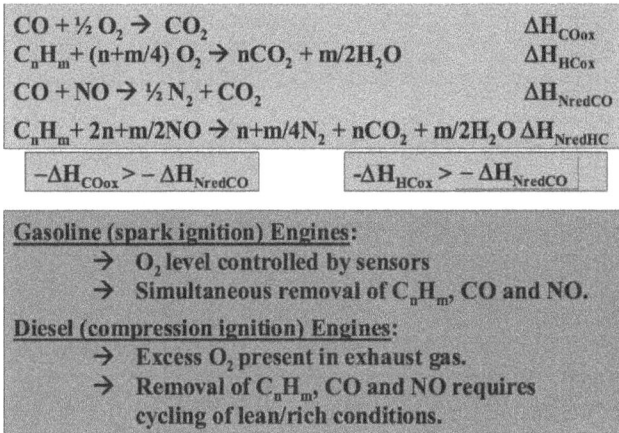

Fig. 6.1.2 Thermodynamic Considerations (Ref. 42)

for oxidation of HC with $O_2(-\Delta H_{HCox})$ is greater than that for the corresponding oxidation with NO $(-\Delta H_{NOredCO})$. Since NO is one of the regulated pollutants, catalysis which utilizes chemical reduction capacity from urea/ammonia, CO and HC is an important technology for meeting clean air requirements.

In gasoline engines, this is accomplished by careful control of the O_2 content by sensors, which maintain O_2 levels very close to stoichiometry. Since no excess O_2 is present, it does not compete with NO for oxidation of CO or HC and thus, NO can be effectively reduced to N_2 in existing "3-way" (NO_x, HC and CO) catalyst systems.

In diesel engines, however, the gas composition cannot be so carefully controlled. Thus, there is always excess O_2 present, making it difficult to reduce NO by reaction with CO or HC. One way this is handled is by modulating back and forth between lean and rich conditions. In the lean condition (low fuel:air ratio), NO is oxidized to NO_2 and stored, while at rich conditions (excess fuel added to the exhaust), HC and CO are oxidized by either the stored NO_2 or by NO.

6.1.3 *Exhaust Gas Components*

The composition of exhaust gas can vary widely based on the type of engine and the conditions. One "typical" set of values (Table 6.1.3) for

Table 6.1.3:   Exhaust Gas Components (Refs. 1, 42a, 43)

Typical Exhaust Gas Compositions		
Component	Gasoline	Diesel
NO_x	1,050 ppm	270 ppm
Soot/PM	30 mg/mi	650 mg/mi
CO	6,800 ppm	350 ppm
Hydrocarbons	750 ppm	270 ppm
H_2	2,300 ppm	120 ppm
O_2	5,100 ppm	60,000 ppm

gasoline and for diesel engines for NO_x, soot (the largest component of particulate matter or PM), CO, HC, H_2 and O_2. Exhaust gas temperatures are typically much higher for gasoline engines, which generate more NO_x and CO, but less soot, consistent with the greater degree of combustion. As was mentioned in 6.1.2, the much higher level of O_2 present in diesel engines makes the simultaneous removal of soot, CO, hydrocarbon and NO_x a thermodynamically challenging problem. As will be discussed, 3-way (hydrocarbon, CO and NO_x removal) catalysts are common in modern gasoline engines, but the analogous situation for diesel engines is not possible because of the high levels of O_2 present. In addition, the high levels of soot (carbon) produced in diesel exhaust further complicates the exhaust thermodynamics and creates the need for additional after-treatment catalysis.

6.1.4 *Catalytic Conversion of Exhaust Gas*

Exhaust gas components can be divided into two categories: oxidizing components and reducing components (Fig. 6.1.4). In the former category are O_2, NO_x and SOx, while the reductants present are CO, soot (mainly C), HC and H_2 (present in very small amounts). In addition, ammonia (NH_3) produced from urea, can be added to the exhaust gas as a reducing component (as a means of removal of NO_x by a process known as selective catalytic reduction or SCR — see Fig. 6.1.8). The catalytic removal or

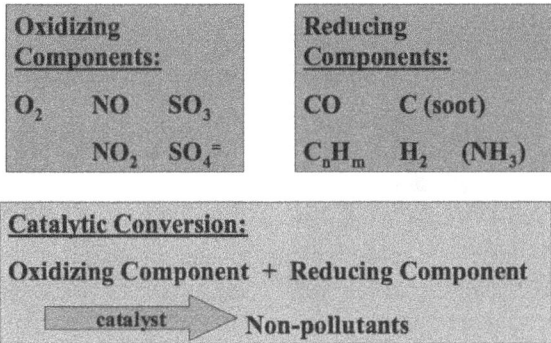

Fig. 6.1.4 Catalytic Conversion of Exhaust Gas (Ref. 42)

"conversion" of these pollutants involves the reaction of oxidizing components and reducing components to form non-pollutant products. The specific reactions by which this occurs is shown and discussed in Fig. 6.1.4 for removal of the major exhaust gas pollutants: NO_x, CO, HC and soot. These are the major "regulated" pollutants, for which legislation limits the level that can be emitted into the atmosphere. For example, EPA requirements for NO_x and PM (soot) mandates their reduction by about 90% in this time period.

6.1.5 *Catalytic Reactions for Removal of Exhaust Pollutants*

NO_x reduction (Fig. 6.1.5.1) can occur by several chemical processes.

The "3-way catalyst" for gasoline engines involves the reaction of CO and NO at low O_2 concentrations to produce N_2 and CO_2. HC can also react with NO_x to form N_2, CO_2 and water by a process known as lean-NO_x catalysis. In the presence of O_2, some of the hydrocarbon is oxidized by oxygen to form additional CO_2 and water. In this case, enough HC must be present to react with nearly all the O_2 that is present so that the remainder can react with the NO_x that is present. Urea can be added to an exhaust to produce NH_3, which reduces NO_x by selective catalytic reduction (SCR). Since most NO_x (90% in diesel exhaust) is present as NO (with 10% NO_2), the reaction is shown for the reduction of NO with ammonia in Fig. 6.1.4.

NO Reduction:

$2NO + 2CO \longrightarrow N_2 + 2CO_2$
(3-way catalyst)

$C_mH_n + (2m + \frac{1}{2}n)NO \longrightarrow (m + \frac{1}{4}n)N_2 + mCO_2 + \frac{1}{2}nH_2O$
(lean NOx catalyst)

$C_mH_n + NO_x + O_2 \longrightarrow N_2 + CO_2 + H_2O$
(lean NO$_x$ catalyst - O$_2$-rich)

$NO + 4NH_3 \longrightarrow 5N_2 + 6H_2O$
(Selective Catalytic Reduction)

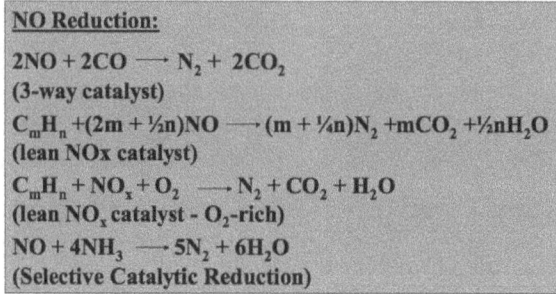

Fig. 6.1.5.1   Catalytic Reactions for Removal of NOx (Refs. 1, 42, 44)

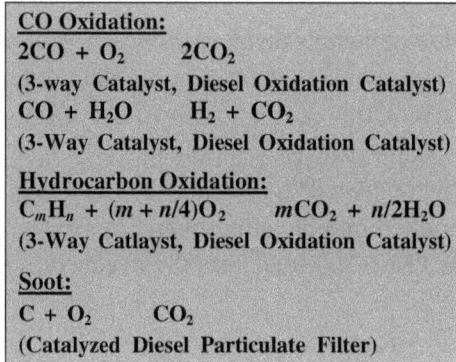

CO Oxidation:
$2CO + O_2 \quad 2CO_2$
(3-way Catalyst, Diesel Oxidation Catalyst)
$CO + H_2O \quad H_2 + CO_2$
(3-Way Catalyst, Diesel Oxidation Catalyst)

Hydrocarbon Oxidation:
$C_mH_n + (m + n/4)O_2 \quad mCO_2 + n/2H_2O$
(3-Way Catlayst, Diesel Oxidation Catalyst)

Soot:
$C + O_2 \quad CO_2$
(Catalyzed Diesel Particulate Filter)

Fig. 6.1.5.2   Catalytic Reactions for Removal of C-containing Exhaust Pollutants (Refs. 1, 42b, 44)

The major process for CO removal is by oxidation to CO_2, which occurs in a 3-way gasoline catalyst, as well as in a diesel oxidation catalyst (DOC) (Fig. 6.1.5.2). A minor reaction which also occurs in these catalysts is the reaction with water to form H_2 and CO_2 (the water-gas shift reaction, discussed in Sec. 5.1.2). Hydrogen is a potent reductant which can be used to reduce NO_x to N_2 and water. HC is also removed by oxidation to CO_2 in a 3-way catalyst. Soot is also removed by its oxidation to CO_2, which occurs on a catalyzed diesel particulate filter (DPF). In this device, soot from the exhaust is first trapped on the catalyst surface which has been treated with a noble metal, and is subsequently oxidized as the temperature in the catalyst reaches the combustion point of the soot (see Sec. 6.1.10).

6.1.6 *Catalysts for Removal of Exhaust Pollutants*

Table 6.1.6   Catalysts for Removal of Exhaust Pollutants (Refs. 1, 42, 44)

	Catalyst Systems
3-Way	$Pt/Rh/LaO/CeO/Al_2O_3$
Lean NOx	$Cu/ZSM5$ (HT) or Pt/Al_2O_3 (LT)
SCR	Pt, V_2O_5/Al_2O_3 or Zeolite
NOx Absorber	$Pt/Rh/BaO/Al_2O_3$
DOC	Pt/Al_2O_3
DPF	noble metal: Pt, Pd, Rh, or Ru, or
	non-Pt-Group: V, Mg, Ca, Sr, Ba, Cu, Ag

The major commercial catalyst systems for the removal of the main exhaust gas components are summarized in 6.1.5.

These are:

- 3-way (gasoline) Catalyst (Gasoline engines),
- Lean NO_x Catalyst (Deisel Engines),
- SCR Catalyst (Diesel Engines),
- NO_x Absorber (Diesel Engines),
- DOC (Diesel engines),
- DPF (Diesel engines).

Each of these will be discussed in detail in the remainder of Sec. 6.1.

6.1.7 *Catalyst Components and Functions — 3-Way Catalyst*

The 3-way catalyst is used in gasoline engines for simultaneous reduction of NO, CO and hydrocarbons (C_mH_n). The components of the catalyst (Fig. 6.1.7.1) are:

- A high surface area γ-alumina which provides the support for the cata-lytic Noble metals Ceria, which acts as an oxygen buffer by storing O_2 when it reaches a certain critical partial pressure,
- Lanthanum oxide, which stabilizes the γ-alumina phase and prevents Rh from dissolving into it,
- Platinum, which is the catalyst for hydrocarbon, CO and NO_2 oxidation,
- Rhodium, which is the catalyst for NO_x decomposition and
- Palladium, which can optionally be used to partially replace Pt or Rh.

2NO + 2CO ⟶ N2 + 2CO2	CO + H2O ⟶ H2 + CO2
2CO + O2 ⟶ 2CO2	CmHn + (m+n/4)O2 ⟶ mCO2 + n/2H2O

• γ-Al2O3 – High surface area support for Noble metals.
• CeOx – Buffers O2 level by O-storage in CeOx matrix.
• LaO – Stabilizes γ-Al2O3 and prevents Rh dissolution.
• Pt – Catalyst for HC, CO and NO oxidation.
• Rh – Catalyst for NOx decomposition.
• Pd (optional): Can partially replace Pt or Rh

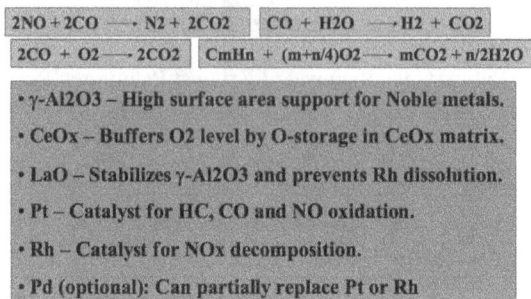

Fig. 6.1.7.1   3-Way Catalyst (Refs. 1, 42, 44)

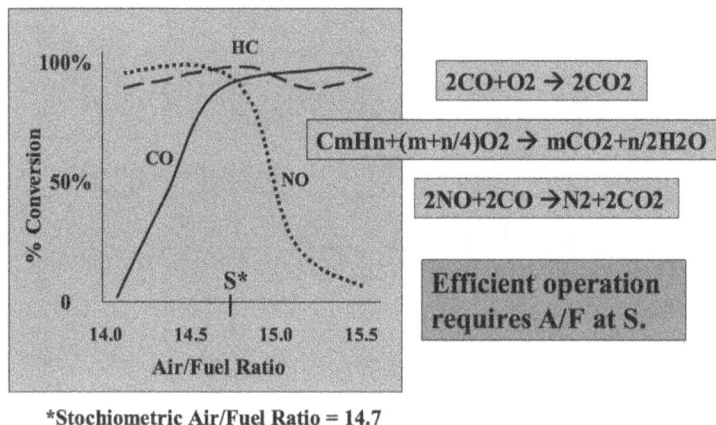

$$2CO + O_2 \rightarrow 2CO_2$$

$$CmHn + (m+n/4)O_2 \rightarrow mCO_2 + n/2H_2O$$

$$2NO + 2CO \rightarrow N_2 + 2CO_2$$

Efficient operation requires A/F at S.

*Stochiometric Air/Fuel Ratio = 14.7

Fig. 6.1.7.2   Simplified 3-Way Catalyst Efficiency as a function of Air:Fuel Ratio (Refs. 1, 44)

The practical operating conditions of this catalyst are shown in more detail in Fig. 6.1.7.2.

The high conversion of CO, NO and hydrocarbon simultaneously is very dependent on maintaining a *stoichiometric* air:fuel ratio S* (i.e., just enough oxygen to totally combust the fuel.) For gasoline engines S* equals a volumetric ratio of 14.7:1. The requirement to maintain air:fuel at S* can be understood by consideration of the three reactions which are utilized in the 3-way catalyst:

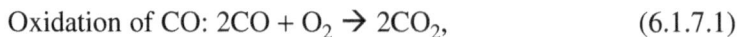

$$\text{Oxidation of CO: } 2CO + O_2 \rightarrow 2CO_2, \qquad (6.1.7.1)$$

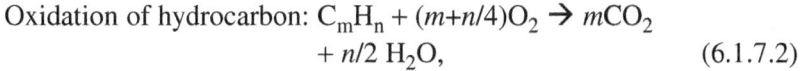

Oxidation of hydrocarbon: $C_mH_n + (m+n/4)O_2 \rightarrow mCO_2$
$+ n/2\ H_2O,$ (6.1.7.2)

Reduction of NO by CO: $2\ NO + 2CO \rightarrow N_2 + 2CO_2.$ (6.1.7.3)

On the *rich* side (i.e., air:fuel ratios < S*), CO conversion falls off because not enough oxygen is present to oxidize it to CO_2. There is much more CO than NO, and so NO alone cannot oxidize all the CO present (by reaction 6.1.7.3). On the *lean* side (i.e., air:fuel ratios > S*), Eq. 6.1.7.1 now effectively competes with Eq. 6.1.7.3 for CO, and so there is less CO available for NO reduction than is required for high conversion of NO. Hydrocarbon conversion by Eq. 6.1.7.2 is fast enough to occur across the range of 14.0< air:fuel <15.5, although it is also highest at S*.

6.1.8 *Catalyst Components and Functions — Lean NO_x Catalyst*

The next catalyst discussed is the lean NO_x catalyst used with Diesel Engines. This system catalyzes the reduction of NO_x by hydrocarbon under both lean and rich conditions by the corresponding reactions:

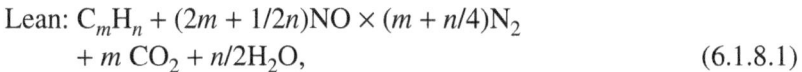

Lean: $C_mH_n + (2m + 1/2n)NO \times (m + n/4)N_2$
$+ m\ CO_2 + n/2H_2O,$ (6.1.8.1)

Rich: $C_mH_n + NO_x + O_2 \rightarrow N_2 + CO_2 + H_2O.$ (6.1.8.2)

The components of this catalyst are designed to allow the catalyst to operate at both low and high temperatures (Fig. 6.1.8). The high temperature

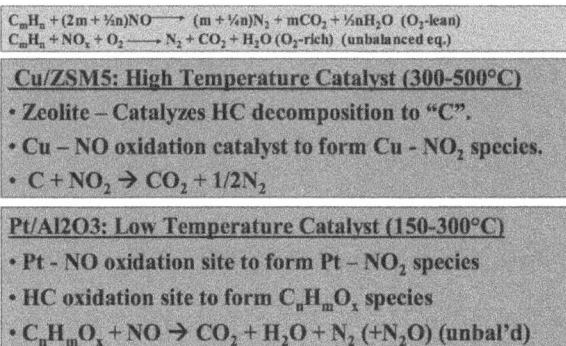

Fig. 6.1.8   Catalyst Components and Functions — Lean NO_x Catalyst (Ref. 42c)

copper on ZSM5 catalyst operates in the range of 300°–500°C and is composed of:

- A **zeolite phase** which catalyzes hydrocarbon decomposition to form active form of carbon,
- **Cu,** which catalyzes the oxidation of NO to NO_2, which in turn is reduced by the active form of carbon to form N_2 and CO_2.

The low temperature platinum on alumina catalyst, which operates at 150°–300°C and is composed of a Pt metal, which catalyzes the oxidation of NO to form a Pt-NO_2 species, and provides a catalytic site for oxidation of hydrocarbon to form an oxygenate species ($C_mH_nO_x$). The oxygenate, in turn, reduces NO to N_2 (and a small amount of N_2O), and, in the process, is oxidized to CO_2 and H_2O. (Ref. 42).

6.1.9 Catalyst Components and Functions — SCR Catalyst

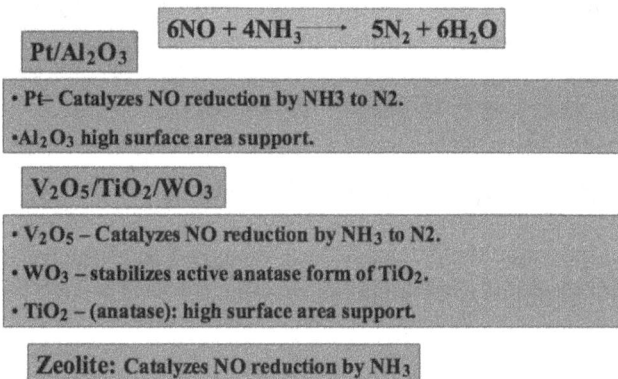

$6NO + 4NH_3 \longrightarrow 5N_2 + 6H_2O$
Pt/Al$_2$O$_3$

- Pt– Catalyzes NO reduction by NH3 to N2.
- Al$_2$O$_3$ high surface area support.

V$_2$O$_5$/TiO$_2$/WO$_3$

- V$_2$O$_5$ – Catalyzes NO reduction by NH$_3$ to N2.
- WO$_3$ – stabilizes active anatase form of TiO$_2$.
- TiO$_2$ – (anatase): high surface area support.

Zeolite: Catalyzes NO reduction by NH$_3$

Selective Catalytic Reduction (or *SCR*) is the process by which ammonia, generated *in situ* from urea hydrolysis, is introduced as a reductant for conversion of NO to N_2 by Eq. 6.1.9 from Diesel engines:

$$6NO + 4NH_3 \rightarrow 5N_2 + 6H_2O. \qquad (6.1.9.1)$$

There are three types of catalysts for this process:

- A noble metal catalyst: Platinum on alumina and
- A base metal catalyst: Vanadia-tungsten oxide on titania
- A zeolite catalyst.

In the noble metal catalyst, the platinum serves as the catalyst for NO reduction by ammonia (i.e., Eq. 6.1.9.1), and the alumina provides a higher surface area support for the Pt catalytic sites.

In the base metal catalyst, vanadia catalyzes Eq. 6.1.9.1, titania (in the alotropic form known as *anatase*) provides a high surface area support, and tungsten oxide stabilizes the high-surface-area anatase phase which is necessary for high conversion.

The zeolite catalysts, which can also contain Fe or Cu, effectively catalyzes Eq. 6.1.9.1.

6.1.10 *Catalyst Components and Functions — NO_x Absorber*

The NO_x absorber is another exhaust after-treatment system for reducing NO_x. The NO_x absorber combines catalytic components for NO_x reduction with bulk NO storage.

Barium oxide is used as the basic NO_x storage media. Under lean conditions (i.e., excess air) NO is first oxidized to NO_2, followed by reaction with barium oxide to form barium nitrate. For a Diesel engine, the exhaust contains excess air, and so the NO_x absorber must remove NO_x under this condition for most of its operation, that is, until all of the available BaO is converted to $Ba(NO_3)_2$. At that time, it must be regenerated under rich conditions, in which the barium nitrate decomposes to NO_x and barium oxide via the reactions shown in Fig. 6.1.10 under "Rich Condition." The NO_x thus formed, as well as the NO_x which is being continuously produced in the exhaust, is reduced by CO to form N_2 and CO_2. The latter reaction,

Table 6.1.10: Catalyst Components and Functions — NO_x Absorber (Ref. 42e)

Lean Condition (O2-Rich):	• BaO – NOx storage
NO + 1/2O2 → NO2	
BaO + NO2 + 1/2O2 → Ba(NO3)2	• Pt – NO oxidation
Rich Condition (O2-Lean):	
Ba(NO3)2 → BaO + 2NO + 3/2O2	• Rh – NO reduction
Ba(NO3)2 → BaO + 2NO2 + 1/2O2	• CeO – O storage
NO + CO → 1/2N2 + CO2	• Al₂O₃ – support

which only occurs under the rich condition, is the basic mechanism that ultimately converts NO_x to a non-pollutant. The rich condition lasts only long enough to regenerate the BaO, which is typically only a fraction of the time at the normal lean condition. After regeneration, the absorber is switched again to the lean condition and the NO_x storage process begins again. This controlled switching between lean and rich conditions is known as *lean-rich modulation*, and as explained above, is required for the operation of a NO_x absorber.

Besides BaO, other components of the NO_x absorber are: Pt, which catalyzes the oxidation of NO to NO_2; Rh, which catalyzes the reduction of NO_x to N_2 by CO, Ce, which buffers the O_2 concentration (as in the 3-way catalyst — Sec. 6.1.7), and a high surface area alumina which functions as a support for the rest of the components.

6.1.11 *Catalyst Functions and Components — DOC and DPF*

The final catalysts discussed in this Section (Fig. 6.1.11) are the DOC, and the DPF. The former uses a platinum-on-alumina catalyst to convert CO to CO_2. The latter traps particulate material (PM), which is mainly soot (carbon), in a ceramic filter, on which is supported a platinum-and-rhodium-on-magnesia catalyst.

The platinum catalyzes the oxidation of the trapped soot, along with any hydrocarbons or CO in the exhaust, to CO_2. The rhodium inhibits the

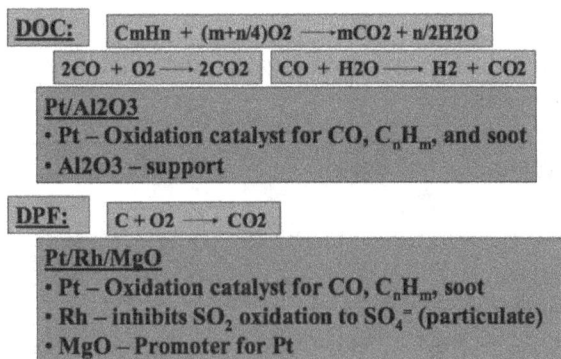

DOC: $CmHn + (m+n/4)O2 \longrightarrow mCO2 + n/2H2O$

$2CO + O2 \longrightarrow 2CO2$ $CO + H2O \longrightarrow H2 + CO2$

Pt/Al2O3
• Pt – Oxidation catalyst for CO, C_nH_m, and soot
• Al2O3 – support

DPF: $C + O2 \longrightarrow CO2$

Pt/Rh/MgO
• Pt – Oxidation catalyst for CO, C_nH_m, soot
• Rh – inhibits SO_2 oxidation to $SO_4^=$ (particulate)
• MgO – Promoter for Pt

Fig. 6.1.11   Catalyst Functions and Components — DOC and DPF (Ref. 42f)

oxidation of SO_2 in the exhaust to sulfate salts, which contribute to particulate matter. The magnesia acts as a promoter for the Pt catalyst.

6.1.12 *Mechanistic Concepts*

As can be seen from the previous discussions in this section, the noble metals (Pt, Pd, Rh) are essential components in a catalytic converter. These metals catalyze several reactions critical for the conversion of pollutants to non-pollutants:

- NO reduction to N_2,
- NO oxidation to NO_2 (for storage as a nitrate),
- Soot (C) oxidation to CO_2,
- Hydrocarbon oxidation to CO_2 and,
- CO oxidation to CO_2.

The mechanisms by which noble metals (M) catalyze these reactions are shown in Fig. 6.1.12.

Platinum is the most common metal present in all major catalytic converters. Platinum can dissociate both CO, to form metal–carbide (M–C) and metal–oxide (M–O) species, and hydrocarbons, to form M–C and

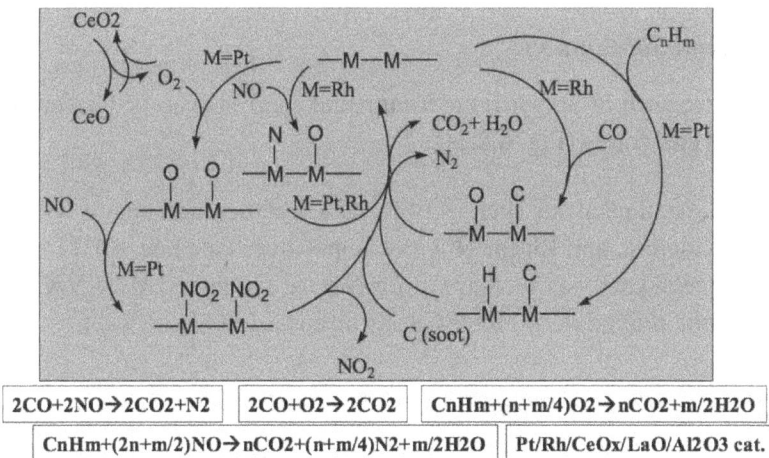

Figure 6.1.12 Mechanistic Concepts (Refs. 1, 42)

M–H species. It can also form metal carbides by reaction with soot (C). Platinum can also dissociate O_2 to form metal oxides, which, in turn can oxidize NO to NO_2 (for storage as $Ba(NO_3)_2$. Palladium can also catalyze these reactions, although it is not as effective as Pt at operating temperatures.

Rh is the other essential noble metal that serves to dissociate NO to metal nitride (M–N) and metal oxide. The rhodium nitride can then react with the platinum (or palladium) carbides or hydrides to form N_2, CO_2 and H_2O. These are the basic mechanisms by which exhaust gas pollutants (NO_x, soot, CO and hydrocarbons) are converted to non-polluting materials (CO_2, N_2 and H_2O.) CeO versibly binds oxygen to acts as an O_2 buffering component.

6.2 Elements of Catalyst Design

As for other catalyst types, the choice of metal, use of promoters, synthetic design and engineering of catalyst form are methods to control the performance of environmental catalysts. An understanding of environmental catalyst poisons is important for the production of long-lived catalysts. Besides specifying a maximum allowable level of pollutants, environmental regulations also require a minimum catalyst life, such as 450,000 miles for catalytic devices on heavy-duty diesel engines.

6.2.1 *Metal Substitution*

As we have seen in Sec. 6.1, environmental catalysts can be chosen from two main categories (Fig. 6.2.1):

- Noble metal catalysts such as Pt for CO, hydrocarbon and
- NO oxidation, and Rh for NO_x decomposition. (see Sec. 6.1.11.)
- Base metal catalysts such as Cu for NO oxidation and V (as V_2O_5) for ammonia reduction of NO (SCR, as discussed in Sec. 6.1.8.)

6.2.2 *Promoters and Poisons*

Promoters are an important means for tailoring the properties of environmental catalysts (Fig. 6.2.2). Some commonly used promoters are MgO

Table 6.2.1:   Metal Substitution (Ref. 42)

Noble Metals:	• Pt: CO, HC, NO oxidation
	• Rh: NO*x* decomposition
Base Metals:	• Cu: NO oxidation
	• V (as V_2O_5): NH_3 reduction of NO

Table 6.2.2:   Promoters and Poisons (Ref. 42)

Promoters:
• For Pt oxidation: MgO
• For Support Phase Stabilization:
  • LaO: Stabilizes Al_2O_3
  • WO3: Stabilizes TiO_2 (anatase)
• Sintering Stabilizers: BaO, La_2O_3, SiO_2 and ZrO_2

Redox Buffering Function:
• CeO: O-storage
• BaO: NO*x* storage

Poisons:
• S, P, Pb, Hg, Cd (for Noble Metals)
• Alkali Metals (for oxidation)
• Thermal deactivation: sintering
• Carrier/catalyst interactions
• Attrition

for Pt oxidation, LaO and WO_3 for stabilization of high surface area γ-Al_2O_3, and the anatase phase of TiO_2 for respectively, and BaO, La_2O_3, SiO_2 and ZrO_2 for stabilization against *sintering* (the agglomeration of solid particles into an amorphous phase which reduces surface area).

Redox buffering functions are an important catalyst component for controlling levels of oxygen and NO_x. The most widely used are CeO for O-storage and BaO for NO_x storage (as nitrate). Several materials can act as poisons for environmental catalysts. Chemical poisons include S, P, Pb,

Hg and Cd for the noble metals, alkali metals for oxidation, and the interaction of catalyst and carrier phases. Physical means of deactivation include thermal deactivation by sintering, and *attrition* (physical loss of fine particles of catalyst.)

6.2.3 *Catalyst Synthesis*

Environmental catalysts are synthesized on a ceramic (*cordierite*) *monolithic* cylinder (or three-dimensional grid), by a 3-step process (Fig. 6.2.3). The cordierite that comprises the monolith cannot be used to support the catalyst because of its low surface area and inability to form highly dispersed metal particles. Therefore, the first step in the process is to adsorb a *washcoat* layer of a stable, high surface area material onto which the catalyst metal particles can be dispersed. Examples of common washcoat layers are shown in Fig. 6.2.3. The final step is a heat treatment of the catalyst to fully disperse the catalyst and to develop the appropriate surface area.

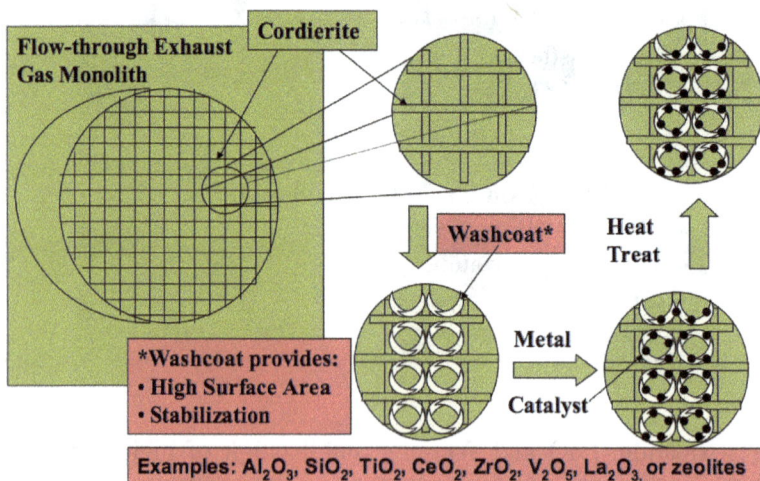

Fig. 6.2.3   Catalyst Synthesis (Ref. 42g)

6.2.4 *Engineering Concepts*

In the design of exhaust treatment catalysts, it is important to understand the factor which limits the efficiency of the process (Fig. 6.2.4). Many of the reactions we have discussed thus far have been chemical-reaction

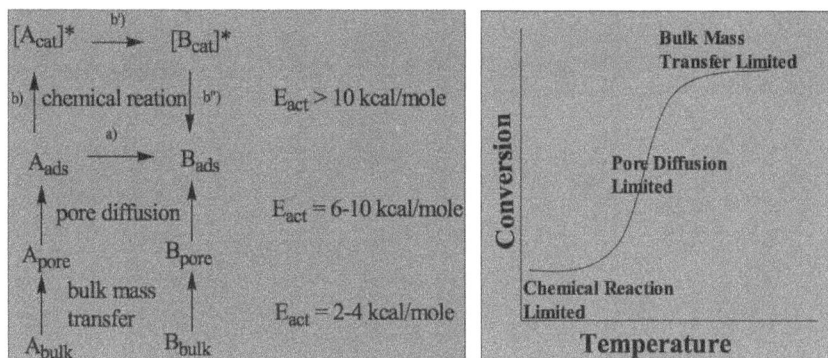

Fig. 6.2.4 Engineering Concepts (Ref. 42h)

limited. Because the reactions in after-treatment devices must occur in a much shorter period of time (on the order of 100 msec) than most catalyst processes for chemicals or fuels production, the temperatures and therefore the conversions per unit time are higher, and the lower activation energy processes become limiting. (The higher the activation energy, the faster the relative rate becomes at high temerature, and so the lower activation energy processes start to become limiting.) As shown in Fig. 6.2.4, this leads to situations which move from the chemical reaction being rate-limited ($E_a > 10$ kcal/mole) to becoming pore-diffusion limited, i.e., diffusion of reactants in and products out of the catalyst pores ($E_a = 6$–10 kcal/mole). As the temperature is further raised, the reactions become bulk-mass-transfer limited ($E_a = 2$–4 kcal/mole), i.e., the reaction rate is as fast as reactants can diffuse to the catalyst surface.

6.3 Major Catalytic Technologies

This section deals with the major commercial after-treatment catalyst technologies. The first four are for diesel applications (6.3.1–6.3.4), the fifth for gasoline engine 3-way catalyst, and the last with catalyst use in fuels cells as a future technology for emissions reduction.

6.3.1 *Diesel Particulate Trap*

First discussed is the diesel particulate trap. In this device, (Fig. 6.3.1.1) the end of every other cell in the monolith is plugged, which forces the gas

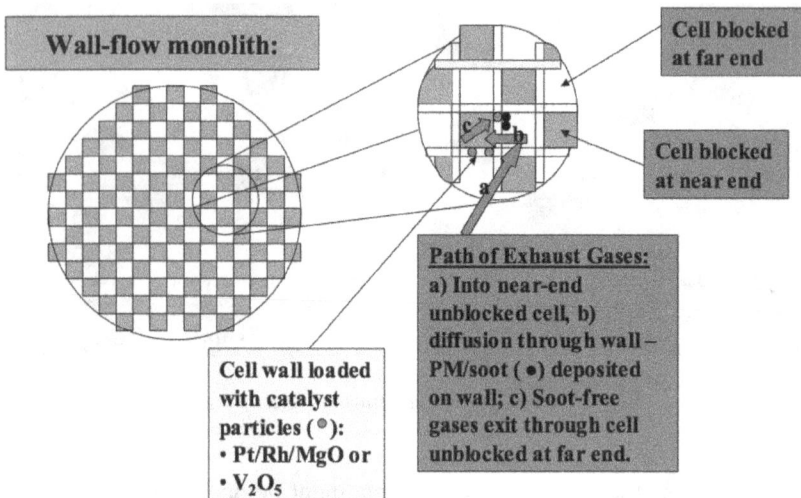

Wall-flow monolith:

Cell blocked at far end

Cell blocked at near end

Cell wall loaded with catalyst particles (°):
• Pt/Rh/MgO or
• V_2O_5

Path of Exhaust Gases:
a) Into near-end unblocked cell, b) diffusion through wall – PM/soot (•) deposited on wall; c) Soot-free gases exit through cell unblocked at far end.

Fig. 6.3.1.1   Diesel Particulate Trap (Ref. 42g)

to pass through the wall of each cell. In the process, the wall acts like a filter, removing soot (along with other particulate matter), which becomes deposited on the cell wall. The cell wall is wash-coated and the wash-coat loaded with particles of a noble metal (Pt/Rh/MgO) or base metal (V_2O_5) catalyst. As the soot builds up on the catalyzed surface and the pores of the ceramic wall become blocked, the back pressure builds up to a level which would start to inhibit the flow of exhaust gases through the filter. At this point, the filter must be regenerated by burning off the soot via catalytic combustion. The process is shown schematically in Fig. 6.3.1.2.

This graph shows the loading of soot and the regeneration of a DPF. The DPF can be operated in either an active or passive mode. In the passive mode, the soot is continuously removed during normal operation. In the active mode, soot is allowed to build up to a certain level, as which time the soot is removed by combustion to regenerate the filter.

In the passive mode, as the level of soot builds up in the filter, the back pressure rises and the soot begins to slowly oxidize, which raises the temperature of the catalyst. When the temperature reaches the "soot burn-off" temperature, the rate of soot combustion becomes very fast and the soot, which is in contact with the combustion catalyst, is rapidly burned to CO_2.

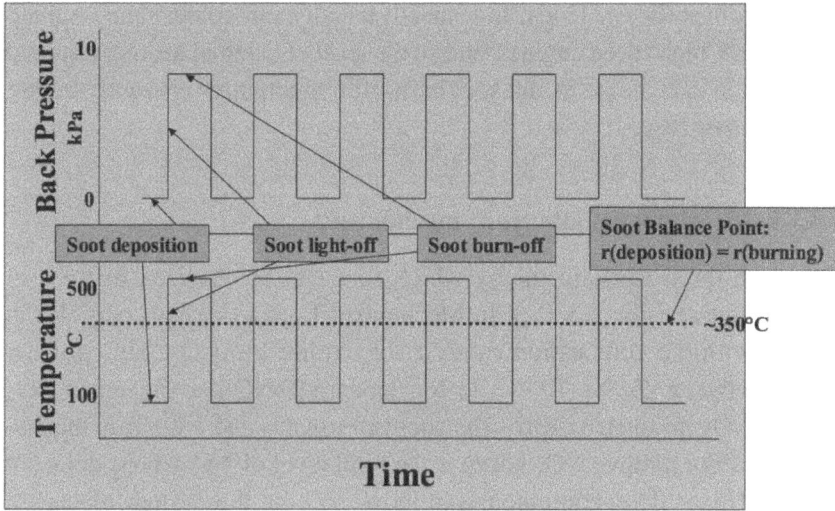

Fig. 6.3.1.2   Diesel Particulate Trap: Balance Point (Ref. 42f)

This process removes soot particles from blocking the pores in the wall, which, in turn, reduces back pressure and restores flow through the filter. The temperature then drops to the normal exhaust temperature because there is no more soot in the filter to burn, and the cycle begins again with soot deposition. The "soot light off temperature" is the point where the soot oxidation rate increases exponentially and all the soot is oxidized.The "soot balance point" is the temperature at which the rate of soot deposition is equal to rate of soot burning. A typical balance point for diesel engines is around 350°C, compared to a soot burn-off point of about 500°C.

In the active mode, soot is allowed to collect until the back pressure reaches a predetermined level. At this point, the temperature is raised to the "soot burn-off" temperature and the filter is regenerated by soot burn-off. As in the passive mode, soot burn-off reduces back pressure and restores flow through the filter. The external heat required can be supplied electrically, or by burning fuel, and so there is a "fuel penalty" associated with the use of these filters, which adds to the operating cost of the vehicle.

The choice of the mode depends on the type of engine, the operating conditions, the levels and types of soot generated, the frequency of regeneration required and the mechanism of regeneration (6.3.1.2). Heavy duty diesel engines (heavy loads, low speed) usually run colder than smaller, lighter load, high speed engines, and so the level of external heating required to attain the balance point and soot burn-off temperatures is usually higher in the former case.

6.3.1.2 Diesel particulate trap: mechanisms

There are three mechanisms by which soot can be oxidized in a diesel particulate trap (Fig. 6.3.1.2). In Mechanism 1, soot (carbon) particles in contact with the combustion catalyst are oxidized to CO_2. This process occurs between 250°C–400°C. In Mechanism 2, NO is oxidized to NO_2 and the NO_2, in turn, oxidizes the soot (C) to CO_2 and NO. This mechanism does not remove NO_x because the total level of $NO + NO_2$ does not change. But it does continuously remove soot as the system reaches a steady state concentration of NO_2. For this reason, Mechanism 2 is known as a continuously regenerated trap or CRT. In Mechanism 3, the uncatalyzed oxidation of soot at 500°C–600°C occurs in "hot spots," which are created from the heat generated by the exothermic reaction: $C + O_2 \rightarrow CO_2$. Since these "hot spots" are not uniform across the catalyst wall, this mechanism is usually minimized to avoid runaway regeneration (uncontrolled temperature rise on the "hot spots,") which can lead to cracking of the ceramic substrate and sintering of the catalyst.

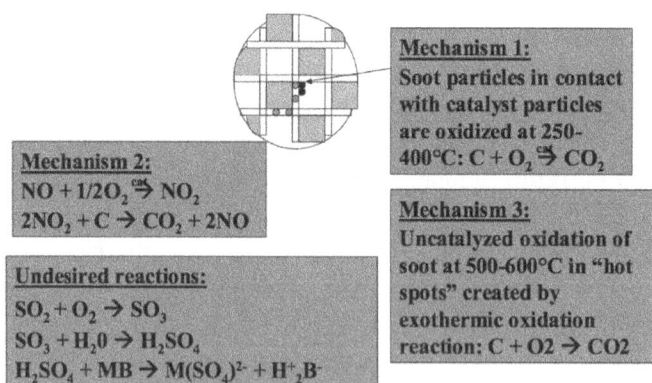

Mechanism 1:
Soot particles in contact with catalyst particles are oxidized at 250-400°C: $C + O_2 \rightarrow CO_2$

Mechanism 2:
$NO + 1/2 O_2 \rightarrow NO_2$
$2NO_2 + C \rightarrow CO_2 + 2NO$

Undesired reactions:
$SO_2 + O_2 \rightarrow SO_3$
$SO_3 + H_2O \rightarrow H_2SO_4$
$H_2SO_4 + MB \rightarrow M(SO_4)^{2-} + H^+{}_2B^-$

Mechanism 3:
Uncatalyzed oxidation of soot at 500-600°C in "hot spots" created by exothermic oxidation reaction: $C + O2 \rightarrow CO2$

Fig. 6.3.1.2   Diesel Particulate Trap (Ref. 42i)

An undesired reaction which occurs on soot oxidation catalyst systems (DPF and DOC's) is the oxidation of SO_2 to SO_3 and subsequent hydrolysis and conversion to sulfate salts. These sulfate salts contribute to particulates and thus, it is important to minimize SO_2 oxidation. The rate of SO_2 oxidation is slower than that of soot oxidation if the noble metal is not poisoned. However, the presence of S in the fuel reduces soot oxidation, making SO_2 oxidation more important (see Fig. 6.3.1.3).

6.3.1.3 Diesel particulate filter: sulfur sensitivity

The noble metal-based DPF systems (Pt and Pd) are the most common for diesel applications because they are more active than base metal systems at the relatively low temperatures of diesel exhaust. These catalysts are, however, very sensitive to sulfur levels in the fuel. Recall that an unwanted process in catalytic oxidation devices (DPF and DOC) is the oxidation of SO_2 to SO_3 and subsequent conversion to sulfate salts (Fig. 6.3.1.3). Since sulfate salts are also particulates, the oxidation of SO_2 to SO_3 in a DFP or DOC actually contributes to particulates. As shown in Fig. 6.3.1.3, the initial conversion of a DPF drops off dramatically as the level of S in the fuel reaches about 50 ppm and higher. In 6.3.1.3, %DPF conversion is about 93–95% at 3 ppm S. This efficiency drops to 72–80% at 30 ppm S.

Sensitivity of a Pt-based DPF to S in Fuel

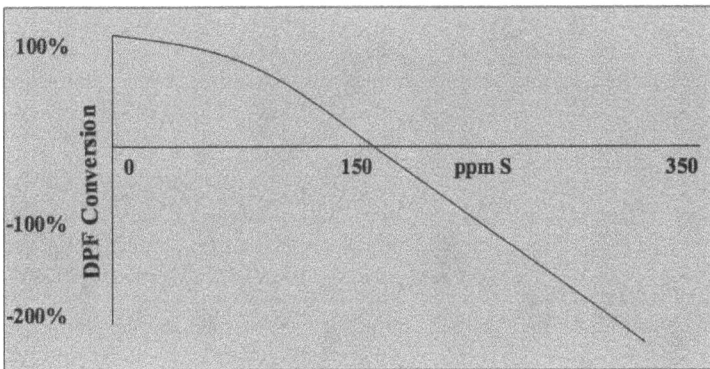

Fig. 6.3.1.3   DPF (Ref. 42j)

At 150 ppm S, the rate of soot oxidation is reduced to such an extent that the particulate formation from SO_2 oxidation and sulfate generation equals the particulate lost by soot oxidation. At S levels > 150 ppm, the DPF becomes a net generator of particulate because the rate of soot oxidation is reduced to less than that of SO_2 oxidation.

Sulfur binds strongly and irreversibly to the surface of the Pt, which blocks the catalytic sites where soot is burned. Since this is a cumulative effect where S builds up on the surface over time, even at very low S concentrations, the entire surface becomes covered with sulfur, leading to deactivation of the catalyst toward soot oxidation.

6.3.2 Diesel Oxidation Catalyst

6.3.2.1 Monolith reactions

The DOC is similar to the DPF, except that the ends of all the channels are open (Fig. 6.3.2.1). As for the DPF, the catalyst is wash-coated with alumina and the catalyst metals are adsorbed on the wash-coated alumina support. The DOC, however, oxidizes soot (C), hydrocarbons and CO continuously as the exhaust gas passes over the walls of the channels, rather than going through soot-collection/burn-off cycles as in the DPF. The reactions which take place in the DOC are the oxidation of CO, soot (C),

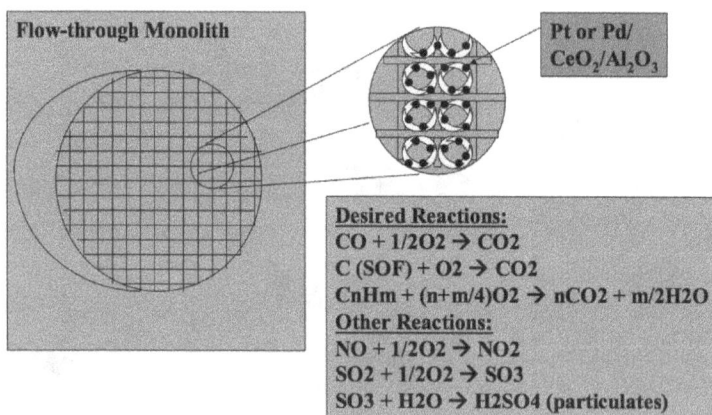

Flow-through Monolith

Pt or Pd/ CeO_2/Al_2O_3

Desired Reactions:
$CO + 1/2 O2 \rightarrow CO2$
$C (SOF) + O2 \rightarrow CO2$
$CnHm + (n+m/4)O2 \rightarrow nCO2 + m/2H2O$
Other Reactions:
$NO + 1/2 O2 \rightarrow NO2$
$SO2 + 1/2 O2 \rightarrow SO3$
$SO3 + H2O \rightarrow H2SO4 \text{ (particulates)}$

Fig. 6.3.2.1 DOC (Ref. 42i)

and hydrocarbons to CO_2. Soot is a heterogeneous material, which is composed of several fractions. Soot results from the dehydrogenation of hydrocarbons by a process that proceeds through intermediates with decreasing H:C ratio. The fraction of these intermediates which are soluble in hydrocarbon solvent is known as the *soluble organic fraction* (or *SOF*), which is composed mainly of heavy polynuclear aromatics. Unlike the DPF which collects all soot and other particulates that are small enough to be trapped on the filter wall, the DOC oxidizes predominantly the soluble organic fraction of soot.

Other reactions that also occur on the DOC, but which do not reduce the overall level of pollutants are the oxidation of NO to NO_2, of SO_2 to SO_3, and the production of sulfuric acid from hydration of SO_3. The sulfuric acid can be neutralize by any basic metals present in the exhaust to form sulfate salts, which contribute to overall level of particulates in the exhaust.

6.3.2.1 Conversion to CO and hydrocarbon

The oxidation (% conversion) of hydrocarbons and CO as a function of temperature for a diesel exhaust is shown in Fig. 6.3.2.2 in the graph on the left for a Pt catalyst. The conversion of CO is greater than that of

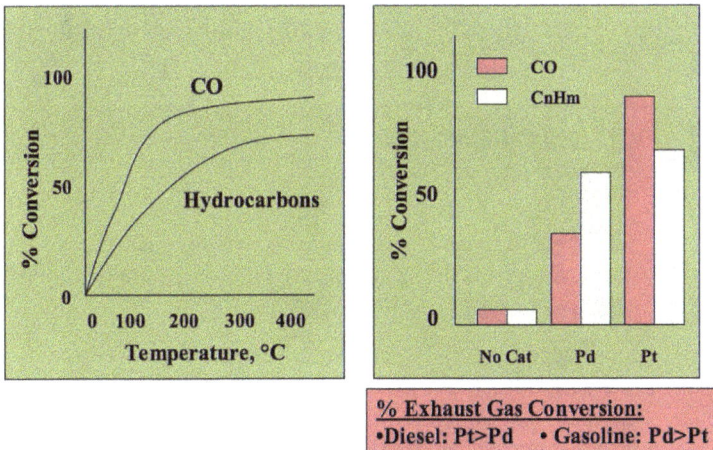

Fig. 6.3.2.2   DOC Efficiencies (Ref. 42b)

hydrocarbons over a Pt catalyst, as shown in the graph on the right. Pt is the catalytic metal of choice for DOC because it is more active for both CO and hydrocarbons than Pd. This is in contrast to gasoline applications, in which Pd is the most active catalyst for hydrocarbons. This is believed to be the result of the shorter chain hydrocarbons in gasoline which react faster over the Pd versus the Pt catalyst, whereas the Pt catalyst is better suited to the longer chains in diesel fuel (Ref. 42).

In addition, Pd is more easily deactivated at high temperatures than is Pt, and therefore, requires more ceria (CeO_x) for thermal stabilization. For these reasons, most modern DOC are Pt-based.

6.3.3 *Diesel* NO_x *Reduction — SCR*

SCR process is the first of several NO_x reduction technologies discussed. As was covered in 6.1.8, SCR is a process by which ammonia is introduced into the exhaust gas as a reductant for conversion of NO to N_2, primarily by the reaction:

$$6NO + 4NH_3 \rightarrow 5N_2 + 6H_2O. \qquad (6.3.3)$$

Other reactions by which NO is reduced by ammonia are shown in Fig. 6.3.3 (under "NO_x-reducing reactions.") Because O_2 is in excess in diesel exhaust, undesirable oxidation reactions can also occur (shown

Table 6.3.3:   SCR Reactions (Ref. 42d)

NO_x-reducing Reactions:
$6NO + 4NH_3 \rightarrow 5N_2 + 6H_2O$
$4NO + 4NH_3 + O_2 \rightarrow 4N_2 + 6H_2O$
$6NO_2 + 8NH_3 \rightarrow 7N_2 + 12H_2O$
$2NO_2 + 4NH_3 + O_2 \rightarrow 3N_2 + 6H_2O$
$NO + NO_2 + 2NH_3 \rightarrow 2N_2 + 3H_2O$
Undesirable Reactions:
$2NH_3 + 2O_2 \rightarrow N_2O + 3H_2O$
$4NH_3 + 3O_2 \rightarrow 2N_2 + 6H_2O$
$4NH_3 + 5O_2 \rightarrow 4NO + 6H_2O$

under "Undesirable Reactions"). These reactions consume ammonia, which makes less of it available for reduction of NO_x. The first and third ones also produce N_2O and NO which actually contributes to NO_x rather than reducing it.

6.3.3.1 SCR mechanism

As was discussed in Sec. 6.1.8, there are three types of catalysts which can by used for SCR, one of which is a base metal catalyst composed of vanadia–tungsten oxide on titania. A mechanism by which NO_x is believed to be reduced to N_2 by NH_3 is shown in Fig. 6.3.3.1. The vanadyl (V=O) groups on the active V(5+) catalyst (V_2O_5) are converted to V=NH by reaction with ammonia and loss of water. The V=NH groups can then react with NO to form a V–NH–N=O species, which is sequentially converted to N_2 and a reduced V (4+) site, which can reduce more NO to N_2, regenerating the active V(5+)=O. The overall reaction in this case is:

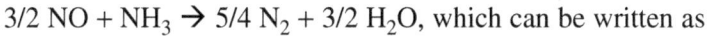

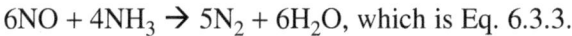

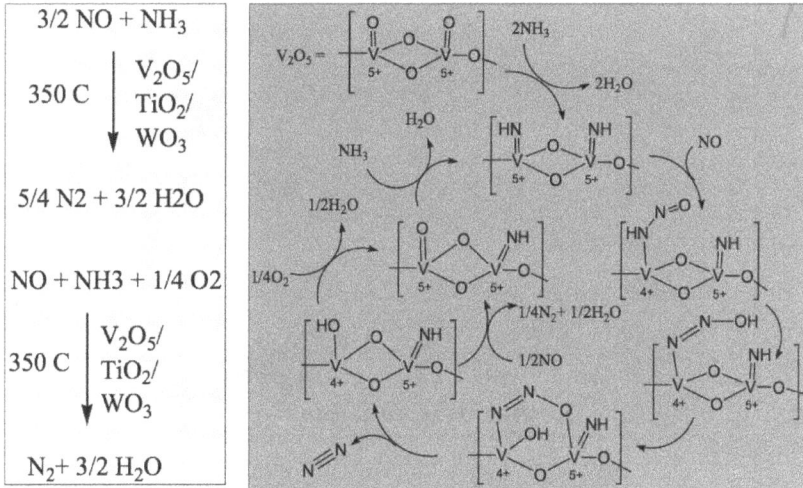

$$3/2 \text{ NO} + NH_3 \rightarrow 5/4 \text{ } N_2 + 3/2 \text{ } H_2O, \text{ which can be written as}$$
$$6\text{NO} + 4NH_3 \rightarrow 5N_2 + 6H_2O, \text{ which is Eq. 6.3.3.}$$

Fig. 6.3.3.1 Proposed SCR Mechanism (Ref. 42d)

Alternatively, O_2 can reoxidize the reduced V (4+) site, in which case the overall reaction is:

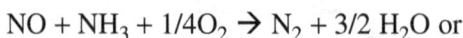

$NO + NH_3 + 1/4O_2 \rightarrow N_2 + 3/2 \ H_2O$ or
$4NO + 4NH_3 + O_2 \rightarrow 4N_2 + 6H_2O$, (Eq. 6.3.3.1, from Table 6.3.3)

which is the second of the NO_x-reducing reaction in Sec. 6.3.3. Both reactions eliminate NO_x and produce non-polluting components (N_2 and H_2O), but Eq. 6.3.3 is more efficient since each mole of NH_3 can reduce one mole of NO. Since there is excess O_2 present in diesel exhaust, this latter process (Eq. 6.3.3.1.) does contribute to the overall conversion of NOx.

6.3.3.2 SCR with V_2O_5/TiO_2 catalyst

The requirement to maintain good control over the ratio of ammonia to NO_x is shown in Fig. 6.3.3.2. If this ratio is too high, unconverted ammonia, which is also a pollutant, will "slip" through into the exhaust and increase the cost of the process by reducing the raw material utilization of NH_3. If the ratio is too low, NO_x conversion will suffer. Thus, it is important to control the ammonia: NO_x ratio to maintain maximum conversion

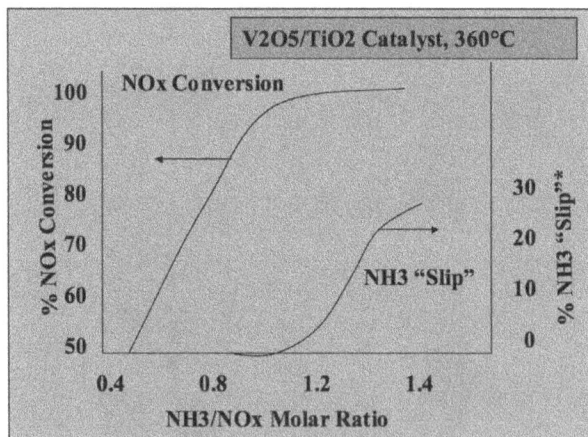

*NH3 "Slip" = % NH3 Unconverted

Fig. 6.3.3.2   SCR (Ref. 42d)

and minimal ammonia "slip." This trade-off is shown in Fig. 6.3.3.2 for the V_2O_5 catalyst system, where the optimal $NH_3:NO$ ratio is around 1.0.

6.3.3.3 SCR: Pt versus V_2O_5/TiO_2 catalysts

The other two catalyst systems used for SCR are the Pt and the zeolite systems (discussed in detail in Sec. 6.1.9). These systems provide NO_x reduction at lower and higher temperatures than the V_2O_5 system, as shown in Fig. 6.3.3.3. The Pt on Al_2O_3 system has a narrow maximum for NO_x conversion at about 150°C, whereas the V_2O_5/TiO_2 and zeolite systems have broader maximum operating ranges at about 300°–400°C and 375°–500°C, respectively. Pt systems are better for heavy duty diesel engines which have a lower exhaust temperature than light duty, higher speed diesel engines.

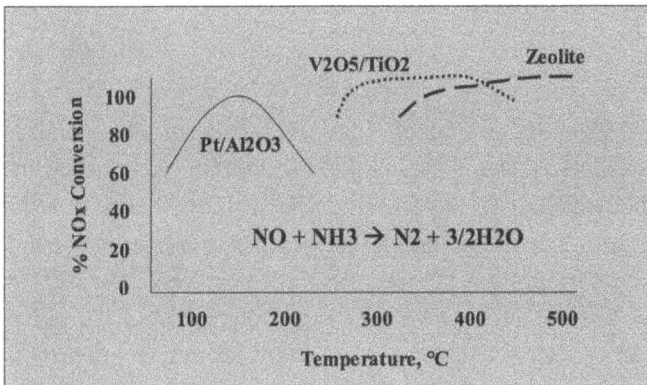

Fig. 6.3.3.3 SCR (Ref. 42d)

6.3.3.4 SCR components

A schematic of a typical SCR vehicle system is shown in Fig. 6.3.3.4. Rather than carrying a pressurized tank of ammonia on board, an aqueous urea solution in a corrosion-resistant tank is used as a source of ammonia. The urea solution is injected into the hot exhaust which hydrolyzes the urea $[(NH_3)_2C=O]$ to NH_3 and CO_2. A DOC can be placed upstream from

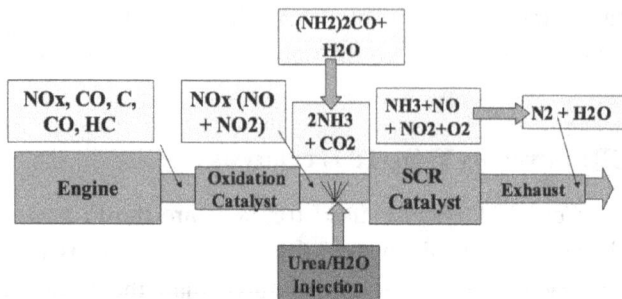

Fig. 6.3.3.4   SCR (Ref. 42d)

the SCR to convert the CO, C (soot) and unburned HC from the engine to CO_2. The urea solution is injected downstream of the DOC, but in front of the SCR catalyst, where the NO_x is reduced to N_2 by the NH_3 formed from the hydrolysis of the urea solution.

6.3.4 NO_x Trap

The next NO_x-reduction technology to be discussed is the NO_x trap (Fig. 6.3.4). This system does not add an external reductant as in the case of SCR, but rather uses the CO and HC in the exhaust to reduce NO_x. However, since this reduction cannot occur in the presence of excess O_2 (i.e., lean condition, as is the case for diesel engines), the NO_x trap functions by storing the NO_x under the lean condition until it can be reduced during a very short, rich cycle that is (artificially) generated by the injection of fuel into the exhaust. Recall the components of the NO_x trap, which are discussed in Sec. 6.1.9 — BaO, Pt, Rh and Al_aO_3. During the lean condition, NO_x is stored as $Ba(NO_3)_2$, by oxidation of NO on the Pt catalyst, followed by reaction of the resulting NO_2 with BaO.

Eventually, all the BaO will be converted to $Ba(NO_3)_2$, at which time the NO_x storage process will not occur and NO_x will begin to appear in the exhaust. Some time before that occurs, the exhaust must be switched to the rich condition (by injection of fuel into the exhaust) for a very short period. During the short rich cycle, the Ba decomposes to BaO, releasing NO_2 and NO, which are immediately reduced to N_2 and CO_2 by hydrocarbon

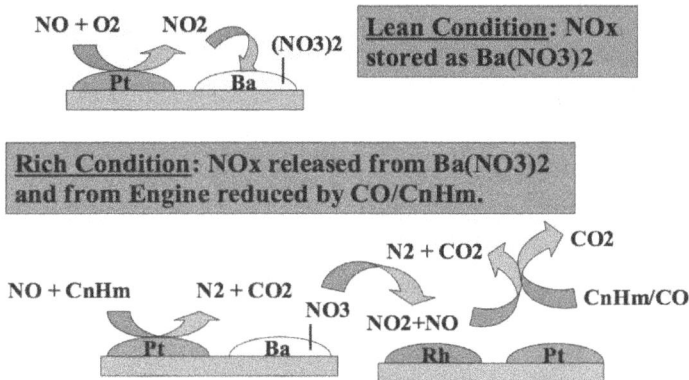

Fig. 6.3.4   NO*x* Trap (Ref. 42e)

(C_nH_m) and CO at the surface of the Rh metal. Pt can also catalyze the reduction of NO by hydrocarbon (C_nH_m) to N_2 and CO_2 (lean NO_x reaction — see Fig. 6.1.7). When all the BaO is regenerated, the exhaust can then be switched back to the normal, lean operating condition. This switching back and forth between a lean and rich condition is known as rich–lean modulation and must be carefully controlled to achieve high, continuous NO_x conversions. This is accomplished by use of gas sensors and electronically controlled injection of fuel into the exhaust. A typical lean-rich cycle might last about one minute, where the engine is running lean for most of that time, with a few seconds of rich regeneration.

6.3.5 *Gasoline Engines — 3-Way Catalyst*

NOx is reduced in a gasoline engine simultaneously along with CO and hydrocarbon using the 3-way catalyst (Sec. 6.1.6). A proposed mechanism for the processes by which this is accomplished on a $Pt/Rh/CeO_x/LaO/Al_2O_3$ catalyst is shown in Fig. 6.3.5. Recall that in a gasoline engine, there is no excess O_2, since this engine runs at a stoichiometric air:fuel ratio, which can be maintained by the $CeO/O_2/CeO_2$ equilibria and controlled by O_2 sensors. Under these conditions, where M=Rh, NO dissociates to form Rh–O and Rh–N species and likewise, CO forms the analogous Rh–O and Rh–C species. These can then combine to form N_2, CO_2 and H_2O.

The Rh–O species can also oxidize soot to CO_2. When M=Pt, it reacts with hydrocarbon to form Pt–C and Pt–H species, which can combine with the Rh-N and Rh–O species to form N_2, CO_2 and H_2O. These reactions are shown at the bottom of Fig. 6.1.12.

Pt also is an oxidation catalyst for NO via formation of Pt–O species by reaction with the small amounts of O_2 that might be present, and subsequent reaction of the Pt–O with NO to form NO_2. The NO_2 thus formed can be reduced to N_2 by CO and hydrocarbon on Rh in an analogous fashion to the reduction of NO.

6.3.6 *Future Technology: Fuels Cells*

An application for environmental catalysis that will find commerical application in the longer term is the fuel cell. Fuel cells convert H_2 and O_2 to H_2O and electricity. There are many types of fuel cells, but the one that is most likely to find use in vehicles is the proton exchange membrane (PEM) cell. This cell generates electricity by reduction of H_2 at the anode to produce protons and electrons which flow in an external circuit (i.e., an electrical current) to the cathode, where O_2 is reduced to O^{2-}. The protons produced at the anode migrate through a polymeric PEM to the cathode, where they combine with the O^{2-} to produce water. Pt on C serves as the catalyst for both the anode and cathode reactions (Fig. 6.3.6).

The advantage of fuel cells is the high energy efficiency that is possible because there are no Carnot cycle limitations of the internal combustion engine. This, in turn, results in dramatically higher theoretical fuel economy compared to the internal combustion engine. In one scenario, the fraction of the propulsion power of the fuel delivered to the wheels is increased to about 35% from about 15% for an internal combustion engine. This is the ultimate in emissions reduction, since the fuel cell produces no NO_x, CO, soot or hydrocarbon emissions, and, because of the higher fuel economy, also produces, on average, less greenhouse gas (CO_2) per mile. While the actual operation of the fuel cell produces no emissions, it must be recognized that the hydrogen used to run fuel cells comes from the steam reforming of hydrocarbons (see Sec. 5.1.5), a process which requires heat

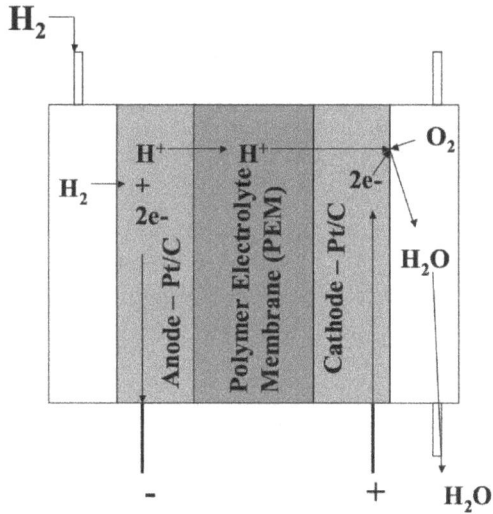

Fig. 6.3.6 **Future Technology: Fuel Cells** (Ref. 45)

from combustion of a hydrocarbon fuel, which produces its own emissions. These emission must be factored into those of a fuel cell operation. This "cradle-to-grave" analysis is important to understand the real impact of new technologies on emissions.

Fuel cells, while they offer a dramatic efficiency increase over the conventional IC engine and make no emissions, they still have major size, weight and cost limitations, which will probably limit their ultimate commercial use versus the internal combustion engine.

6.3.6.1 Fuel cell reactions

The mechanism by which these processes occur is shown in Fig. 6.3.6.1. At the cathode, Pt is oxidized to Pt–O, which combines with the two electrons from the external circuit, and two protons (H+) that had been transported to the cathode by the protonated membrane (H+–Mem–H+), to produce water. At the anode, Pt dissociates H_2 to form a Pt–H species, which in turn, loses two electrons, and releasing two protons to the PEM for transport to the cathode.

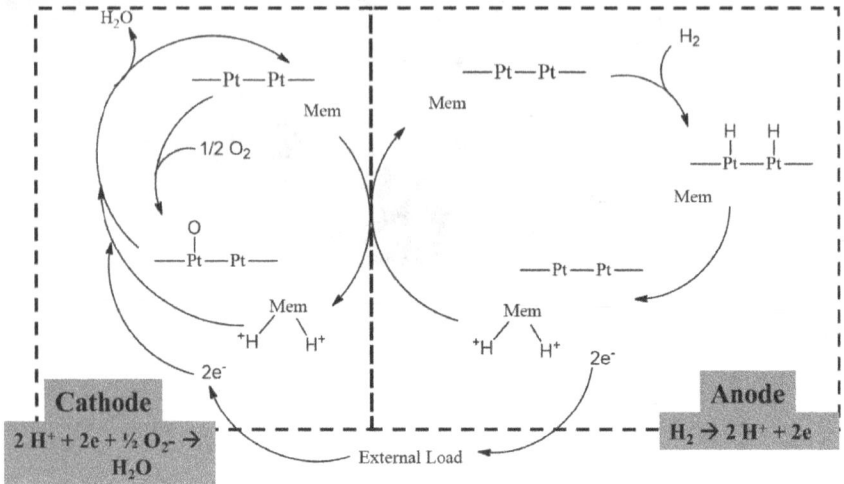

Fig. 6.3.6.1   Fuel Cell Reactions (Ref. 45)

6.4 Problems

1. Match a process from Column A with a reaction from Column B and a catalyst from Column C:

A	B	C
(a) Lean NOx	(1) $C + O_2 \rightarrow CO_2$	(i) Pt/Rh/LaO/ CeO/Al$_2$O$_3$
(b) 3-way catalyst	(2) (a) $CO + O_2 \rightarrow CO_2$ (b) $C_nH_m + O_2 \rightarrow CO_2 + H_2O$	(ii) Pt/Al$_2$O$_3$
(c) SCR	(3) $NO + O_2 \rightarrow NO$ $NO_2 + BaO + O_2 \rightarrow Ba(NO_3)_2$ $Ba(NO_3)_2 \rightarrow NO + O_2 + BaO$ $NO + CO \rightarrow N_2 + CO_2$	(iii) BaO/Pt/Rh/ CeO/Al$_2$O$_3$
(d) NOx Absorber	(4) $C_nH_m + NO \rightarrow N_2 + CO_2$ $C_nH_m + NO + O_2 \rightarrow N_2 + CO_2 + H_2O$	(iv) Pt/Rh/MgO
(e) DOC	(5) (a) $CO + \frac{1}{2}O_2 \rightarrow CO_2$ (b) $NO + CO \rightarrow \frac{1}{2}N_2 + CO_2$ (c) $C_nH_m + O_2 \rightarrow CO_2 + H_2O$	
(f) DPF	(6) $6NO + 4NH_3 \rightarrow 5N_2 + 6H_2O$	

2. Match the catalytic element (A) with the corresponding function (B):

A	B
(a) Pt	(i) NOx storage metal
(b) Rh	(ii) high surface area support
(c) CeO	(iii) NO dissociation
(d) γ-Al$_2$O$_3$	(iv) O$_2$ buffer
(e) TiO$_2$ (anatase)	(v) stabilizes γ–Al$_2$O$_3$
(f) WO$_3$	(vi) catalyst for CO, NO and C$_n$H$_m$ to active "C"
(g) BaO	(vii) catalyzes decomposition of C$_n$H$_m$ to active "C"
(h) MgO	(viii) stabilizes TiO$_2$ (anatase)
(i) Zeolite	(ix) catalyzes NO reduction by NH$_3$ to give N$_2$
(j) V$_2$O$_5$	(x) Promoter for Pt
(k) LaO	

3. Write the metal Pt and/or Rh over each reaction step in which it is involved in the following mechanism:

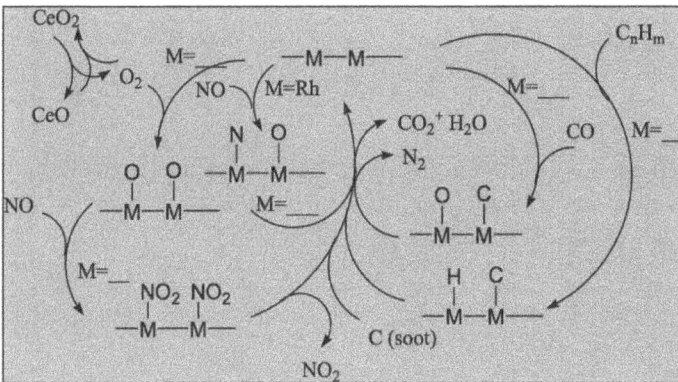

4. Using the graph below, explain why the 3-way catalyst must operate under tight control at the stoichiometric air:fuel ratio (14.7) with no excess oxygen.

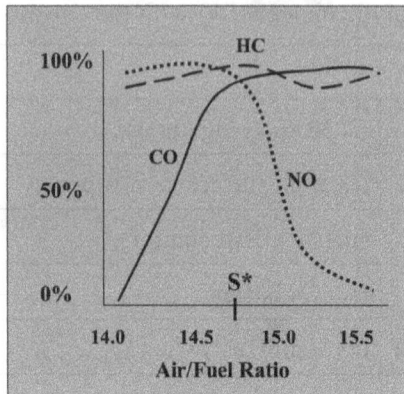

5. Identify the components of the NOx absorber below as Ba or Rh:

Lean Condition:

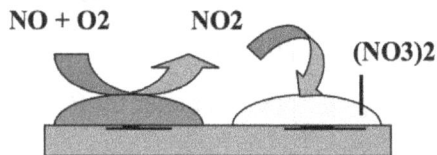

Rich Condition:

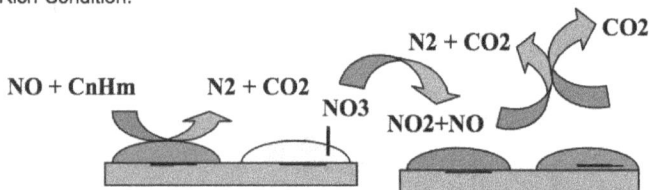

6. Explain how electricity is generated from a hydrogen fuel cell by labeling the following on the figure below:

Anode (Pt)

PEM

Cathode (Pt)

H_2

O_2

H_2O

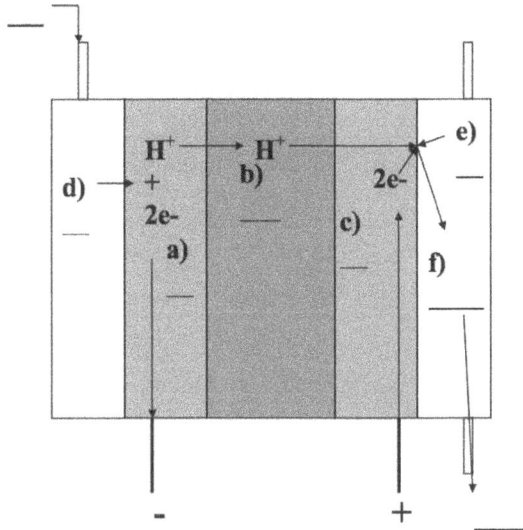

6.5 Answers to Problems

1.

A	B	C
(a) Lean NOx	(4) (a) $C_nH_m + NO \rightarrow N_2 + CO_2$ (b) $C_nH_m + NO + O_2 \rightarrow N_2 +$ $CO_2 + H_2O$	(ii) Pt/Al_2O_3
(b) 3-way catalyst	(5) (a) $CO + 1/2 O_2 \rightarrow CO_2$ (b) $NO + CO \rightarrow 1/2 N_2 + CO_2$ (c) $C_nH_m + O_2 \rightarrow CO_2 + H_2O$	(i) $Pt/Rh/LaO/$ CeO/Al_2O_3
(c) SCR	(6) $6NO + 4NH_3 \rightarrow 5N_2 + 6 H_2O$	(ii) Pt/Al_2O_3
(d) NOx Absorber	(3) (a) $NO + O_2 \rightarrow NO_2$ (b) $NO_2 + BaO + O_2 \rightarrow Ba(NO_3)_2$ (c) $Ba(NO_3)_2 \rightarrow NO + O_2 + BaO$ (d) $NO + CO \rightarrow N_2 + CO_2$	(iii) $BaO/Pt/Rh/$ CeO/Al_2O_3
(e) DOC	(2) (a) $CO + O_2 \rightarrow CO_2$ (b) $C_nH_m + O_2 \rightarrow CO_2 + H_2O$	(ii) Pt/Al_2O_3
(f) DPF	(1) $C + O_2 \rightarrow CO_2$	(iv) $Pt/Rh/MgO$

2.

A	B
(a) Pt	(vi) catalyst for CO, NO and CnHm oxidation
(b) Rh	(iii) NO dissociation
(c) CeO	(iv) O_2 buffer
(d) γ–Al_2O_3	(ii) high surface area support
(e) TiO_2 (anatase)	(ii) high surface area support
(f) WO_3	(viii) stabilizes TiO_2 (anatase)
(g) BaO	(i) NOx storage material
(h) MgO	(x) Promoter for Pt
(i) Zeolite	(vii) catalyzes decomposition of CnHm to active "C"
(j) V_2O_5	(ix) catalyzes NO reduction by NH_3 to give N_2
(k) LaO	(v) stabilizes γ–Al_2O_3

3.

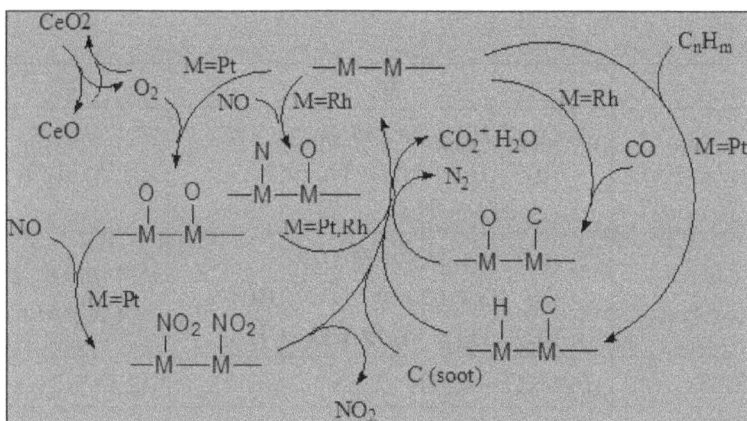

4.

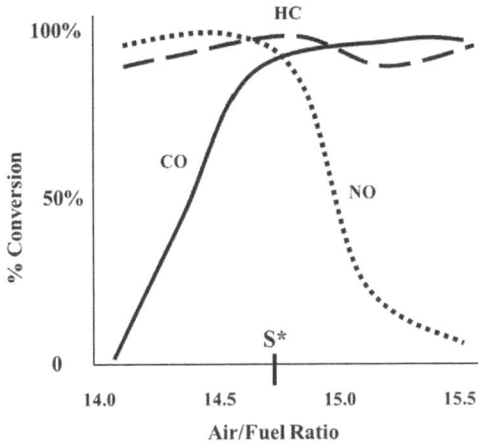

On the rich side (air/fuel < 14.7), there is not enough O_2 to oxidize all the CO (CO + ½ O_2 → CO_2), but there is more than enough to reduce NO (NO + CO → ½ N_2 + CO_2). On the lean side (air/fuel>14.7), there is not enough CO to reduce NO (NO + CO → ½ N_2 + CO_2) because there is so much O_2 that all the CO is used up to form CO_2 (CO + ½ O_2 → CO_2).

5. Lean Condition

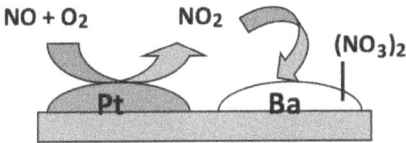

Rich Condition:

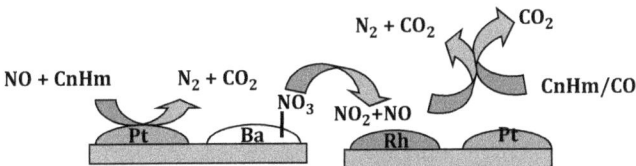

7

Catalyst Characterization

In this final chapter are highlighted several of the methods by which catalysts are characterized. These are divided into two types: spectroscopic and adsorption methods (Fig. 7).

Spectroscopic methods use some form of electromagnetic energy, in the form of radiation (photons) or energetic particles, as a probe for a chemical or structural feature of the catalyst or bulk. These features are derived from measurement of the way in which the energetics or particles of the probe is modified by the catalyst. Commonly used probes are X-rays, infrared radiation, radio waves and electrons. As for other analytical methods, those for catalyst characterization have their own set of acronyms (shown in Fig. 7.1), by which the methods are more commonly known.

Another type of catalyst characterization is the measurement of the adsorption and/or the desorption of molecular species from the surface of a catalyst (Fig. 7). The amount and nature of the molecules which are adsorbing and desorbing gives information regarding the physical nature of the bulk and surface of the catalyst, such as surface area, pore volume and pore size. The use of an adsorbing probe molecule on a catalyst surface and subsequent measurement of the resulting chemical species and heat released or absorbed as a function of temperature thermogravimetric analysis (TGA) and differential scanning calorimetry (DSC) can provide important mechanistic information on the chemisorption, desorption and the nature of the intermediates involved in a catalytic reaction.

238 Industrial Catalysis: Chemistry and Mechanism

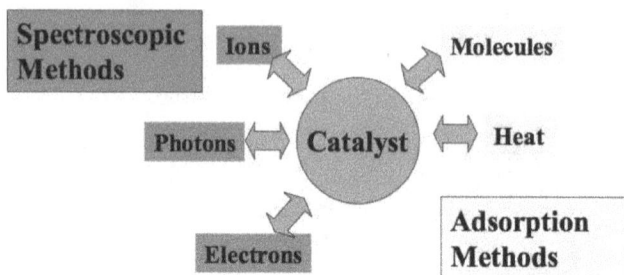

Fig. 7 Catalyst Characterization (Ref. 1, 40)

7.1 Spectroscopic Methods

The various probes and resulting measurements for spectroscopic and adsorption methods are illustrated in Fig. 7.1. Spectroscopic methods use ions, photons and electrons as probes and measured species, while adsorption methods most commonly utilize and measure heat and the desorption of molecules. The various spectroscopic techniques for catalyst characterization are discussed in terms of the corresponding combinations of these probes and measured species.

Several techniques will be discussed that use photons as probes (photons in) and measure photons (photons out.) These are:

- X-ray diffraction (XRD) and X-ray fluorescence (XRF) spectroscopy
- Extended X-ray adsorption fine structure (EXAFS)
- Fourier transform infrared (FTIR) spectroscopy
- Raman spectroscopy
- Nuclear magnetic resonance (NMR) spectroscopy

One technique that uses photons as probes (photons in) and measures electrons (electrons out) — X-ray photoelectron spectroscopy (XPS) — will be discussed. Several techniques that use electrons as probes and measure electrons are presented:

- Scanning electron microscopy (SEM)
- Transmission electron microscopy (TEM)
- Scanning TEM (STEM)
- Scanning tunneling microscopy (STM)
- Auger microscopy

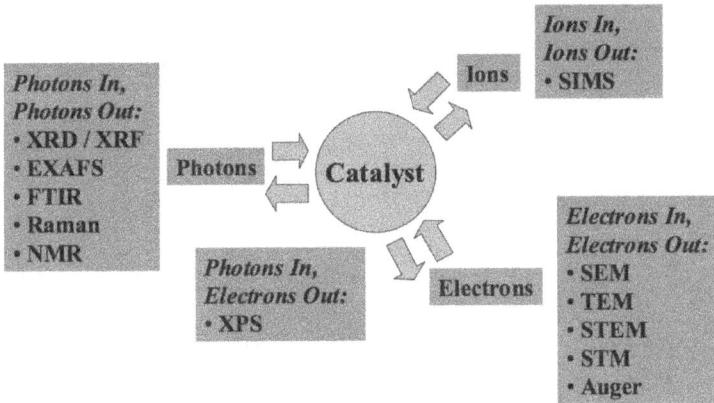

Fig. 7.1 Spectroscopic Methods (Ref. 40)

Secondary ion mass spectroscopy (SIMS) is an ions in/ions out method that is also discussed in this chapter.

Table 7.1 summarizes the important features of each of the spectroscopic methods to be discussed in Chapter 7:

Energy In: The electromagnetic radiation or energetic species that is used to probe the catalyst.

Energy Out: The electromagnetic radiation or energetic species that is emitted from and which measures a chemical or structural aspect of the catalyst.

Sample Crystallinity: The requirement of the method for a single crystal, or whether a polycrystalline sample (i.e., powder with grain boundaries), or an amorphous sample (i.e., no long range order) can be used.

Measurement: The actual measurement that is made which correlates to some chemical or structural feature of the catalyst.

In Situ: The method can be adapted to analyze samples under actual catalytic operating conditions.

Ex situ: The method cannot be used under catalytic conditions, but requires a separate set of conditions (e.g., high vacuum or low temperature) under which the catalyst would not be expected to operate.

Analytical Information: The chemical and/or structural features which the method provides.

Table 7.1:   Spectroscopic Methods (Refs. 1, 40)

Method	Energy In	Energy Out	Sample	Measurement	In/Ex-Situ	Information	Bulk / Surface	Sample required	Quati-tation
XRD/XRF	hv - X-rays	hv - X-rays	Polycrystalline	Angle (2θ, d) & intensity of diffracted X-rays or fluorescence	In situ; ex situ (limted by high energy beam)	Crystallographic phases & crystallinity	Bulk	mg's	Rel. % phases
XPS	hv - X-rays	electrons (eV)	Polycrystalline or amorphous	Binding energy of emitted photoelectrons	In situ (limted by high energy beam)	Elements & oxidation states	Surface	mg's	Rel. % elements
EXAFS	hv - X-rays	hv - X-rays	Polycrystalline	Intensity & energy of adsorption edge X-rays	In situ (limted by high energy beam)	Coordination number (N) and bond distances	Bulk	mg's	N & d (known structures)
FTIR	hv - IR	hv - IR	Single crystal (r-a); polycrystalline (disp.)	IR spectrum of adsorbed species	In situ (reflection-adsorption); ex situ (dispersive)	Nature & location of adsorbed species	Surface	g's	Rel. mol. concentra-tions
Raman	hv - IR	hv - IR	Polycrystalline or amorphous (non-fluorescent)	Raman spectrum of adsorbed species	In situ and ex situ	Nature & location of adsorbed species	Surface	g's	Rel. mol. concentra-tions
NMR	hv - radio	hv - radio	Polycrystalline or amorphous	NMR spectrrum of solid catalyst or adsorbed species	Ex situ (instument configuration limitations)	Catalyst elements/ox. states; adsorbed molecular structures	Bulk (solid); surface (ads. species)	g's	Rel. mol. concentra-tions
EM/EDAX	electrons	electrons	Polycrystalline	Electron diffraction pattern/energy of emitted X-rays	Semi ex situ (high vacuum limitations)	Crystallographic phases, crystallinity, crystal size	Bulk	mg's	Rel % phases; % crystal.; part. size
SIMS	ions	ions	Polycrystalline or amorphous	Mass/charge ratio of emitted secondary ions	Ex situ (high vacuum limitations)	Catalyst surface elements/clusters	Surface	µg's	Not quanti-tiative

Bulk versus Surface: Whether the method provides information about the surface (i.e., first few atomic layers) of the catalyst, or the catalyst bulk (i.e., the average of all the atoms in a solid catalyst.)

Amount of Sample Required: An order of magnitude estimate of the amount of sample Required.

Quantitation: The aspects of the analytical information which can be quantified, i.e., the aspects about which one can identify how much is present in the catalyst.

7.1.1   X-ray Diffraction/X-ray Fluorescence

The first method discussed is XRD and XRF. In this method, a polycrys-talline sample is placed in the path of an X-ray beam in which the angle from the plane of the surface is varied by an instrument known as an X-ray diffractometer. The intensity (I) of the diffracted X-rays (XRD) or fluores-cence (XRF) is measured as a function of the angle of the beam from the

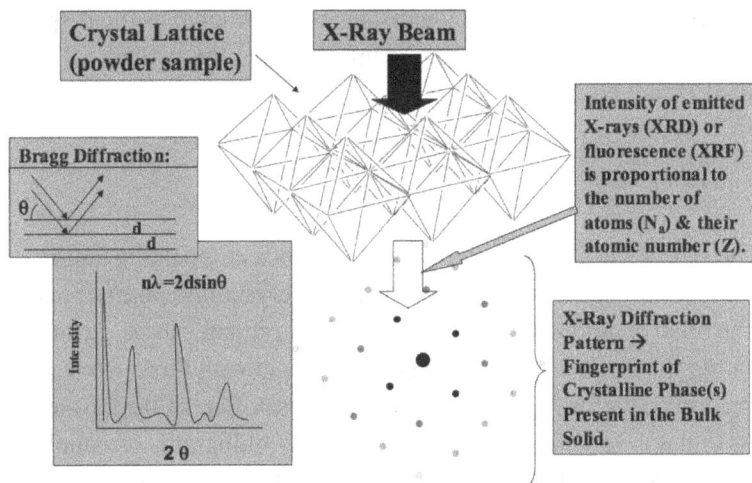

Fig. 7.1.1 XRD/XRF (Refs. 1, 40)

surface (Θ). The intensity of the emitted X-rays or the resulting fluorescence (XRF) is proportional to the number of atoms (N) and their atomic number (Z). Thus, the diffraction pattern provides a blueprint of the crystalline phases (i.e., material of like composition and long-range order) present in the bulk solid (Fig. 7.1.1).

The XRD or XRF spectra are reported as intensity of the X-ray radiation as a function of 2θ (Fig. 7.1.1, lower left corner). This graph is derived from the diffraction pattern (Fig. 7.1.3, bottom center). This pattern is a series of dark spots of varying intensities, resulting from diffraction of the X-rays by atoms in consecutive layers of the bulk. The intensity of the spots decrease as the beam has to travel through more layers of repeating atomic units.

7.1.1 XRD/XRF

- Crystalline Phases Present
  — from computer search of phase data library
  — Phase composition
  — XRF is very sensitive (0.1–1 monolayer of atoms) and requires only milligrams of sample
- Crystallinity of the Catalyst
  — Sharp peaks → crystalline (long-range order)
  — Broad or no peaks → amorphous
- Size of Microcrystallites

If each layer is separated by a distance d (as known as the d-spacing), and the wave length of the X-ray beam is λ, then the pattern follows the so-called Bragg equation:

$$n\lambda = 2d\sin\theta,$$

where n is an integer.

The size of the d-spacing reflects the distance of repeating units (or unit cell) in the solid and is a characteristic of the crystalline phases present.

The relative amounts of crystalline phases present can be quantified by comparing the XRD or XRF spectra of a sample with those from a library of diffraction patterns of known crystalline phases. One of the advantages of XRF is that it is very sensitive and requires only milligrams of sample to get a diffraction pattern. Another is that multiple XRF samples can be analyzed at the same time using automated XRF batch analyzers. XRD and XRF methods depend on the presence of long range order in the bulk solid. A diffraction pattern with highly defined long range ordering (i.e., highly crystalline) will have very sharp diffraction peaks. A diffraction pattern will not be produced at all if there is no long range ordering, i.e., if the sample is totally amorphous. Samples which have some short-, medium-range ordering of atoms (several repeat units, but not throughout the entire sample), or which have mixtures of crystalline and amorphous regions will give rise to broad diffraction peaks. Based on an analysis of the size and breadth of diffraction patterns of a known crystalline phase, the size of the microcrystallites (i.e., size of the crystals between grain boundaries) can be determined.

7.1.2 *X-ray Photoelectron Spectroscopy*

The next catalyst analytical method to be discussed is X-ray photoelectron spectroscopy (XPS, Fig. 7.1.2). Like XRD and XRF, the catalyst sample to be analyzed is placed in the path of an X-ray beam, but instead of recording a diffraction pattern, photoelectrons which are emitted from the sample detected. (Other electrons such as scattered electrons and *Auger* electrons are also emitted and can be detected to give information about the solid, but these methods will not be discussed in detail in this Chapter.)

The electrons associated with a given atom are strongly bound to the nucleus of that atom by an energy equal to the binding energy of those

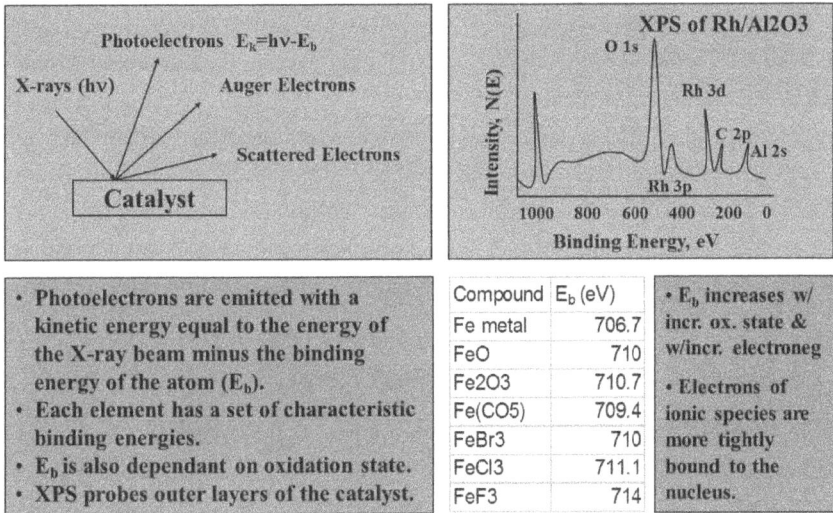

Fig. 7.1.2   XPS (Refs. 1, 40)

electrons to that atomic nucleus. The binding energy is a function of the elemental nucleus and the orbital in which the electron resides. X-rays are very high energy electromagnetic radiation (on the order of a thousand electron volts), which exceeds this binding energy. When an X-ray beam hits a solid surface, the atoms in the surface absorb the energy of the beam and emit an electron. The photoelectrons which are emitted have a kinetic energy equal to the energy of the incoming X-ray beam minus the binding energy of the atom from which the electron was emitted. Each element has a characteristic binding energy for electrons in each of its atomic orbitals. The binding energy is also dependent on the oxidation state of the atom. Because the photoelectrons emitted come only from the first few atomic layers, the elemental information from the XPS method reflects only the elements on the surface of the catalyst, not the bulk average. An XPS spectrum of a Rh on alumina catalyst sample (upper right, Fig. 7.1.2) shows binding energies corresponding to the elements present and the electronic orbitals from which the electrons were emitted. It also shows that a carbon impurity is present on the surface of the catalyst. The intensity of the peaks in this spectra are proportional to the atomic % of the nuclei present. Thus, EPS results are normally tabulated in the form of atomic %s of the various nuclei.

The table of binding energies for various iron compounds (lower right of Fig. 7.1.2) shows that binding energy increases with increasing oxidation state and with increasing electronegativity of the bonding atoms. As a nucleus becomes more positively charged (which occurs with increasing oxidation state and increasing electronegativity of the bonded atoms), the electrons have a greater attraction for the nucleus and therefore the binding energy increases. By this logic, atoms which have a formal positive charge have higher binding energies than the corresponding neutral atom.

7.1.3 *Extended X-ray Adsorption Fine Structure*

In the method known as *EXAFS*, high energy X-rays beyond the adsorption edge of the normal XRD experiment (30–1200 eV), are constructively and destructively backscattered (like ripples from a pebble thrown into a pond) by nearest neighbor atoms to produce an interference pattern known as the EXAFS spectrum (Fig. 7.1.3). The frequency of the pattern is dictated by the interatomic distances (in Å) and the amplitude by the number of nearest neighbor atoms (N) around the central atom. A representative EXAFS spectrum is shown in the center of 7.1.3.

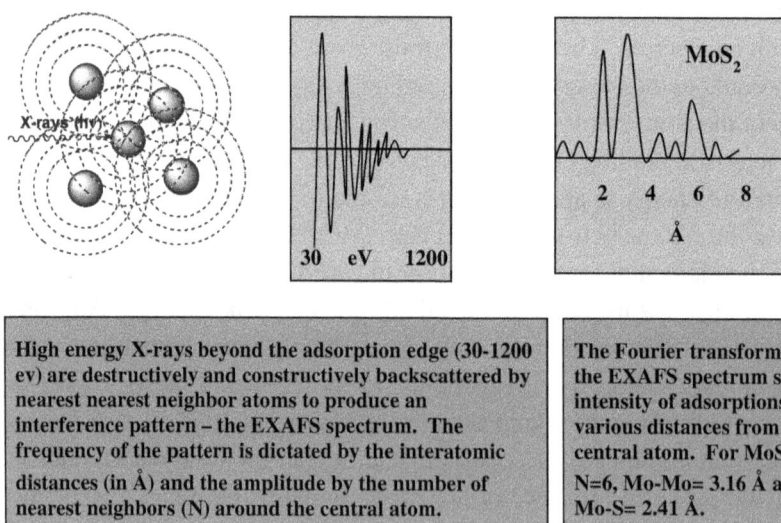

High energy X-rays beyond the adsorption edge (30-1200 ev) are destructively and constructively backscattered by nearest nearest neighbor atoms to produce an interference pattern – the EXAFS spectrum. The frequency of the pattern is dictated by the interatomic distances (in Å) and the amplitude by the number of nearest neighbors (N) around the central atom.	The Fourier transform of the EXAFS spectrum shows intensity of adsorptions at various distances from the central atom. For MoS_2, N=6, Mo-Mo= 3.16 Å and Mo-S= 2.41 Å.

Fig. 7.1.3   EXAFS (Refs. 1, 40)

The EXAFS spectrum is a composite of all the interference effects from a distribution of energies from the diffraction pattern resulting from a composite of all the nearest neighbor distances from the central atom. The "wiggles" in the raw EXAFS spectrum can be deconvoluted to provide the contribution of nearest neighbors at each of the interatomic distances by a computational technique known as Fourier transform. The Fourier transform of an EXAFS spectrum of a solid catalyst of formula MoS_2 (right side of Fig. 7.1.3) shows the intensity of adsorptions at various distances from the central atom. In MoS_2, the coordination number is 6. The nearest neighbors to a central Mo atom are S (Mo–S = 2.41 Å). The other peaks in the Fourier transform spectrum represent the next nearest neighbors at 3.98 Å (Mo–S), 5.47 Å (Mo–Mo) and 6.32 Å (Mo–Mo), respectively.

7.1.4 *Fourier Transform Infrared Spectroscopy*

FTIR (Fig. 7.1.4) also exposes a catalyst sample to multiple wavelengths simultaneously, in this case, with radiation in the infrared region. A spectra results from the deconvolution of absorption at various wavelengths to produce a Fourier, the deconvolution of absorption at various wavelengths to produce a transform spectrum.

In the transmission mode, a thin catalyst sample is irradiated and the adsorption as a function of wavelength is obtained. This is the most straightforward way of obtaining an IR spectra, but if the catalyst sample is opaque or cannot be prepared as a thin sample, as is most commonly the case, a transmission spectra cannot be obtained. An FT diffuse reflectance spectra can be obtained on such samples because it records adsorption of light reflected from the surface of a solid catalyst sample. Because of this mechanism of adsorption, it is very surface sensitive. This is a very versatile technique which can be used on a wide variety of solid catalyst samples and conditions, including catalytic reaction conditions. For example, the method has been used to study adsorbed species and their reactions on catalytic surfaces. Most notable among the adsorbed species studied is that of CO on metals. The various modes of adsorption, as shown in the upper right of Fig. 7.1.4, can be distinguished based on their infrared adsorption in the diffuse reflectance mode. In the example in the

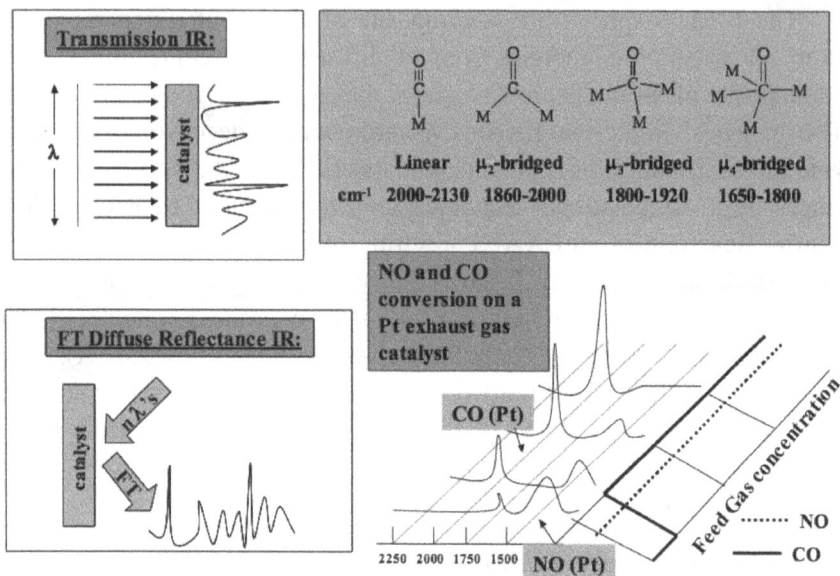

Fig. 7.1.4 FTIR (Ref. 40)

bottom right of Fig. 7.1.4, NO is seen to disappear as CO is introduced into the gaseous feed to a Pt-based exhaust gas catalyst, but the reduction of NO by CO according to equation $CO + NO \rightarrow CO_2 + \frac{1}{2} N_2$ (see Eq. 6.1.6).

7.1.5 *Laser Raman Spectroscopy*

Laser Raman spectroscopy is similar in principle to FTIR, except that the Raman activity generally resides in molecules with low dipole moments. This is complementary to infrared active groups, which relies on a change in dipole moment. For example, IR-active solids such as alumina, silica and titania all have M–O bonds with strong dipole moments, but are transparent to Raman spectroscopy. Raman active groups include oxides of Mo, V and Ni, and molecules containing S–S bonds (such as organic and disulfides). These molecules are IR inactive because these functional groups have a very low dipole moment. Like FTIR diffuse reflectance, laser Raman is a surface-sensitive technique. It is applicable to a wide range of catalyst sample types — crystalline, amorphous or supported and is well suited to *in situ* studies over a wide range of temperatures and pressures.

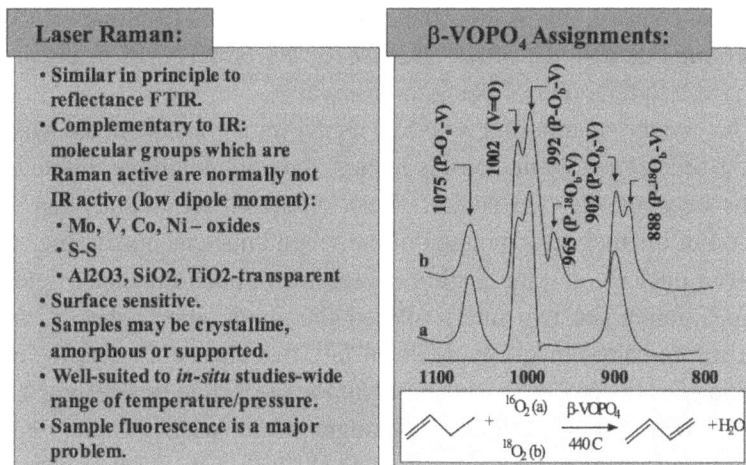

Fig. 7.1.5   Laser Raman Spectroscopy (Ref. 46)

One of the drawbacks is that if a sample has any fluorescence, this will overwhelm any Raman signal, making it impossible to obtain a spectrum.

An example of an *in situ* Raman study is shown on the right side of Fig. 7.1.5 for the oxydehydrogenation of butene to butadiene over a VO (PO_4) catalyst. The spectrum for the catalyst when it is reduced by butene (forming butadiene) and reoxidized by $^{16}O_2$ is shown in Fig. 7.1.5 as (a). The same process using $^{18}O_2$ is shown in (b) where the shifts in the spectrum are consistent with incorporation of ^{18}O into the $V-O_b$ bridging site. This is consistent with the assignment of $V-O_b$ as the oxygen insertion species (see 3.3.6).

7.1.6 *Nuclear Magnetic Resonance Spectroscopy*

NMR spectroscopy is one of the most widely used techniques by chemists because of the wealth of information that it can provide about the nature and environment of chemical species. NMR is normally performed on sample which are dissolved in a liquid solvent. In such samples, the tumbling motion of the molecules averages out the line-broadening interactions between nuclei, known as *anisotropic* interactions. The "averaging-out" of these interactions results in very sharply defined, narrow adsorption peaks. In contrast, for solids, the nuclei are static, resulting in strong *anisotropic* interactions, which give rise to very broad NMR peaks, sometimes, even to the point of being unobservable.

This line broadening for an NMR signal for a given nucleus, which is due to a second nearby nucleus, is in large part, due to dipole-dipole interactions between the two nuclei. These dipole-dipole interactions are a function of the magnetic moment of the second nucleus, its distance from the given nucleus, and the angle between the applied magnetic field and the line joining the two nuclei. When this angle approaches a critical value, the so-called magic angle of 54°54', the term containing the angle function goes to zero, as does the magnitude of the line broadening. NMR on solids with spinning at this magic angle is known as *magic angle spinning nuclear magnetic resonance* or *MASNMR* (Fig. 7.1.6.1), which is now commonly used to obtain high resolution spectra from solid samples. Two other techniques known as decoupling and cross-polarization are frequently used with solid state NMR for signal enhancement. Decoupling is accomplished by irradiating nuclei which are coupled to the one of interest, thereby reducing any splitting to a single peak, which enhances the signal. Line broadening can also be reduced by decreasing the relaxation time of the nucleus of interest. (NMR signals are very broad for nuclei with very long relaxation times.) Cross polarization enhances signal by providing a fast pathway for relaxation of a relatively rare nuclei (most commonly ^{13}C) by relaxation through a more abundant nuclei (typically ^{1}H). For Fourier transform NMR (FTNMR), when the spins

---

Solid State NMR (MASNMR)

- Normally broad lines in solids can be sharpened by magic angle spinning of the sample.
- Magic angle = spinning about an axis 54°54' to the magnetic field
- ^{1}H and ^{13}C analysis of adsorbed molecules
- Analysis of catalytic elements: ^{29}Si and ^{27}Al
- Chemical shift → hybridization and electron density
- Multiplicity → coupling of probe nucleus to nearest neighbors

Fig. 7.1.6.1   NMR (Ref. 40)

relax faster, more FT pulses can be collected in a given period of time, thus reducing the time to obtain a spectrum.

MASNMR can be obtained for adsorbed molecules (most commonly ^{1}H or ^{13}C NMR), or the catalytic elements themselves (such as ^{29}Si or ^{27}Al NMR). The two main properties obtained from NMR spectra, namely chemical shift and multiplicity (splitting) provide information on hybridization/electron density, and coupling of the nucleus of interest to its nearest neighbors.

As an example of the use of MASNMR for catalytic applications, the thermal activation/dealumination of ammonium zeolites can be followed by NMR (Fig. 7.1.6.2). With increasing conversion, the number of aluminums bonded though a bridging atom to tetrahedral silica decreases (left spectra, a → d). Likewise, as aluminum is ejected from the tetrahedral silica phase and a separate octahedral alumina phase forms, the number of aluminum atoms in an octahedral environment increases (right spectra, a → d). (The notation in the silica spectra on the left shows in parentheses the number of Al atoms bonded to Si through bridging O atoms.) Thus, the NMR analysis of silica–aluminas is an important way to determine the extent of dealumination and the nature of the residual aluminum in the silica matrix.

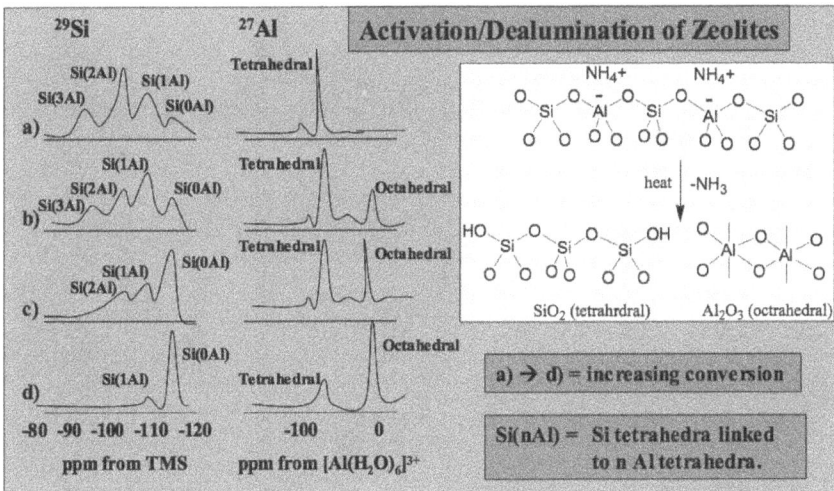

Fig. 7.1.6.2   NMR (Ref. 40)

7.1.7 *Electron Microscopy*

Electron microscopy (EM) is a very versatile technique from which much crystallographic, structural and chemical information can be obtained. In an electron microscope, a thin catalyst sample is bombarded with mono-energetic electrons of 20–1000eV and the resulting diffracted, and transmitted electrons are analyzed (Fig. 7.1.7.1). These electrons can form a diffraction pattern, similar in principle to an XRD pattern (see Fig. 7.1.3.2), from which crystallographic phase and unit cell information can be obtained. Transmitted electrons form an image with atomic resolution, corresponding to magnification of 10^6 to 10^8 times. TEM is very similar in operation to an optical microscope with electrons instead of visible light. Like the optical microscope, TEM provides a two-dimensional image of the sample. The image results from electrons which are slightly diffracted from the transmitted beam. Catalyst applications of TEM include the imaging of metals on a supported catalyst. In SEM, a narrow electron beam is rastered over the surface of the sample and secondary electrons (from the surface) and back-scattered electrons (from deeper in the sample) are detected. Because atoms both at the surface and below the surface can be detected, SEM images show three-dimensional topography of the

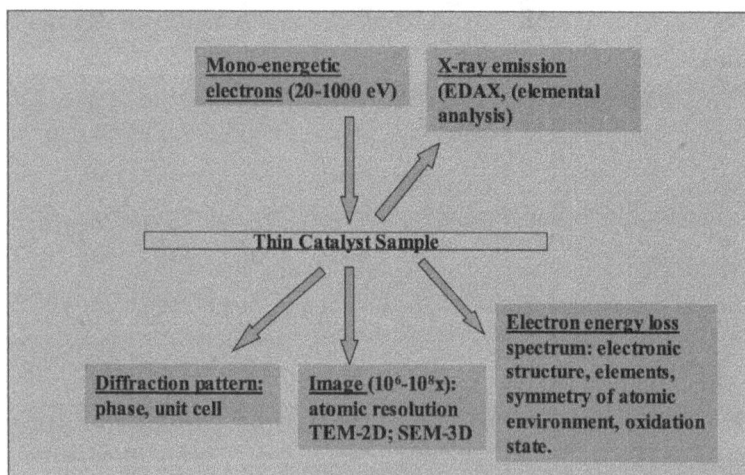

7.1.7.1   EM (Refs. 1, 40)

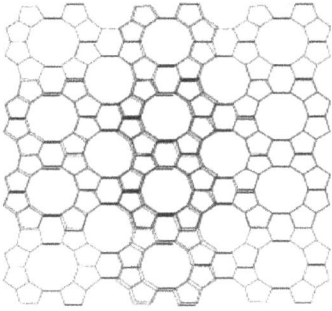

ZSM-5 Zeolite Structure ZSM-5 High Resolution TEM (simulation)

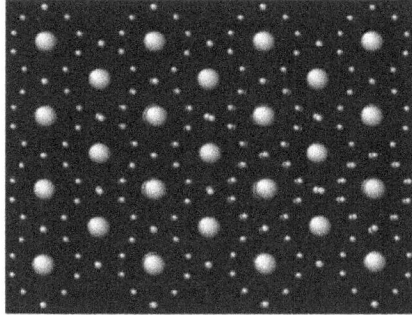

7.1.7.2 Electron Micrographs (Ref. 3)

surface. Electron microscopes which combine both modes are known as STEM instruments.

Electron energy loss spectrum contains information on electronic structure, the elements present and their oxidation state, and the symmetry of the atomic environment. A solid which is bombarded with an electron beam also emits X-rays, which are characteristic of a particular element. The electron beam can be moved across the sample to determine the composition of various regions of the sample. This technique is referred to as electron microprobe analysis or energy dispersive analysis of X-rays (EDAX), which refers to the mode of detection. Many electron microscopes are equipped with EDAX units to allow determination of elemental composition for a selected region of the EM image.

As an example of a catalyst analysis using EM, a high resolution electron micrograph (TEM) of a ZSM-5 zeolite is shown by the simulation in Fig. 7.1.7.2. For comparison, the corresponding structure shown on the left is a projection looking down one crystallographic direction [010].

7.1.8 *Secondary Ion Mass Spectrometry*

SIMS is a highly surface-sensitive probe, in which the surface is bombarded with high-energy particles (usually Ar^+). These ions can reach subsurface atoms of the solid sample and when the energy of the particle dissipates in this subsurface region, a fraction of the energy reaches the

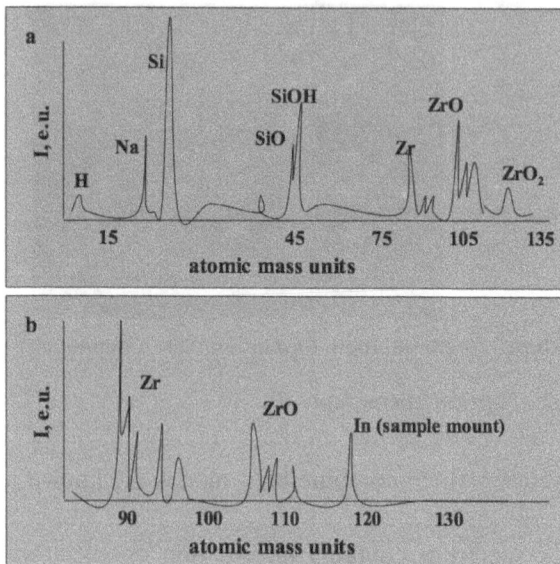

Secondary Ion Mass Spectra of a Zr/SiO2 catalyst a) as synthesized, and b) after drying and calcination. The spectra show dispersion of Zr on the surface of the silica support with heat treatment. SIMS is highly surface sensitive but it is not quantitative.

Fig. 7.1.8   SIMS (Ref. 1)

surface, resulting in the ejection of a "secondary" particle, which can be detected by a mass spectrometer. These secondary particles can be charged or uncharged atoms, clusters of atoms or molecules. While SIMS is extremely sensitive (10^{-5} monolayers of surface atoms), it is not quantitative because the number of secondary ions is not proportional to the surface concentration of the originating atom or molecular species.

The SIMS spectra of a Zr/SiO_2 catalyst (Fig. 7.1.8) shows how the Zr becomes dispersed on the surface as the sample is calcined. Silica, which is present on the surface of the catalyst in the slurry as it is synthesized, is covered by a dispersed layer of ZrO_2 in the calcined material.

7.2 Adsorption Methods

Adsorption methods can be used to determine surface area, pore volume and pore size distribution. The controlled thermal desorption of chemical species from a catalyst can be followed by weight loss TGA or by energy loss or uptake DSC which gives valuable information on the nature of the species present and how strongly it is bound to the surface.

7.2.1 *BET Surface Area*

As we have discussed throughout this course, surface area is a very important parameter in heterogeneous catalysis. Surface area can be defined as the amount of an adsorbing molecule (usually N_2) that is in a filled molecular surface layer or *monolayer*. For most catalyst applications, the BET method (named after the developers — Brunauer, Emmett and Teller) relies on the adsorptive behavior of N_2, as shown on the left graph in Fig. 7.2.1. This involves multilayer adsorption and capillary condensation in *mesopores* of 2–50 nm in diameter. As the pressure of N_2 is increased, the amount adsorbed reaches a plateau, then rapidly increases to another plateau, like a titration curve. From the volume of N_2 adsorbed per gram of catalyst (V_m) at 77°K, the number of N_2 molecules in a monolayer and the corresponding surface area can be calculated using the equations shown in the bottom right of Fig. 7.2.1. B is the inflection point shown on the graph at the left on Fig. 7.2.1, and when divided by the number of grams of sample, is the volume of a monolayer of N_2 molecules per gram of catalyst. The number of N_2 molecules in a monolayer (n_m) is derived from ideal gas law ($0.0224 \ m^3g^{-1}$). The specific surface area (SA in m^2g^{-1}) is the total number of molecules of N_2 (moles times 6×10^{23} molecules/mole) times the average area occupied by an N_2 molecule (a_m).

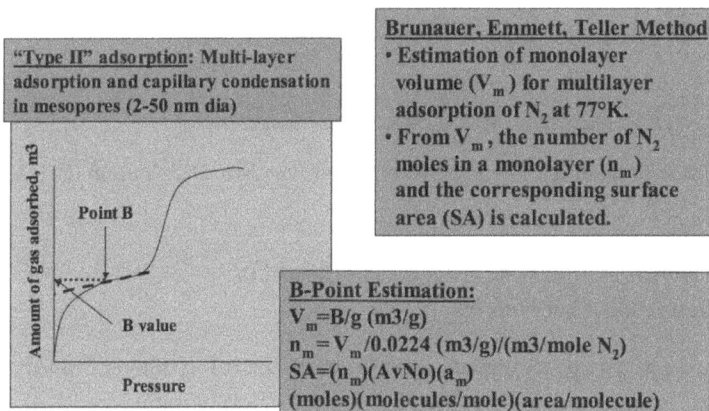

Fig. 7.2.1 BET Surface Area (Ref. 1, 40)

7.2.2 *Porosimetry — Pore Volume and Pore Size Distribution*

A technique called mercury porosimetry is used to determine the total pore volume and the distribution of pore radii in a solid sample with pore sizes in the range of 3.5–75,000 nm (75 microns). As the applied pressure of mercury is varied from about 0.1–2,000 atmospheres, mercury is forced into the pores of the sample. The total volume of mercury adsorbed is inversely proportional to the applied pressure, according to the Washburn equation (Fig 7.2.2). The total volume adsorbed, as pressure is increased, asymptotically approaches the *total pore volume* of the sample (*adsorption curve* — black line, Fig. 7.2.2). The first derivative curve of the adsorption curve is the *pore size distribution* (blue curve, Fig. 7.2.2).

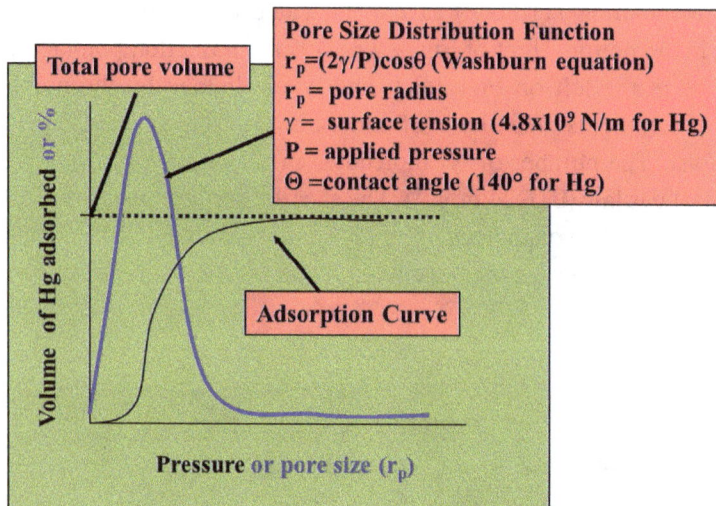

Total pore volume

Pore Size Distribution Function
$r_p = (2\gamma/P)\cos\theta$ (Washburn equation)
r_p = pore radius
γ = surface tension (4.8×10^9 N/m for Hg)
P = applied pressure
Θ = contact angle (140° for Hg)

Adsorption Curve

Volume of Hg adsorbed or %

Pressure or pore size (r_p)

Fig. 7.2.2   Porosimetry — Pore Volume and Pore Size Distribution (Ref. 1, 40)

7.2.3 *Thermogravimetric Analysis/Differential Scanning Calorimetry*

Thermal analytical techniques are a very useful means of analysis of adsorbed species and the modes of decomposition of catalysts. TGA measures the loss of mass from a catalyst as a function of temperature. As the catalyst is heated, the sample will lose weight as any adsorbed molecules desorb, with the more weakly bound coming off at lower temperatures, and the more strongly bonded ones at higher temperatures. As the temperature is further raised to the catalyst's decomposition temperature, further weight loss occurs as products of decomposition are released from the sample. When run under an inert atmosphere, this weight loss corresponds to loss of thermal decomposition products, while under an air or other oxidizing atmosphere, oxidative decomposition products form (Table 7.2.3).

The oxidative decomposition is usually lower than the thermal decomposition temperature. TGA can also be used for determining the thermal behavior of poisoning species which, when desorbed, result in increase in catalytic activity.

DSC involves a similar experimental setup, but instead of weight loss, heat released (exotherm) or adsorbed (endotherm) is measured. Reactions of the solid itself that can also be followed by DCS, including oxidation/reduction reactions, other chemical reactions and phase changes. Like TGA, DSC can also be used to provide information on the desorption and reaction of adsorbed species. TGA and DSC can be done simultaneously, so that a weight loss peak can be correlated with the corresponding endothermic or exothermic process.

In Fig. 7.2.3 is an example of a TGA and the corresponding DSC, run under air, for a catalyst with two adsorbed species (I and II), which desorb at 202°C and 405°C, respectively, and an oxidative decomposition point of 605°C. The TGA shows that the two adsorbed species make up 35% and 15% of the weight, respectively. The active catalyst, which

Table 7.2.3:   TGA/DSC

TGA	DSC
• Measures loss of mass from a catalyst as a function of temperature.	• Measures exo- or endo-thermic processes that catalyst undergoes as a function of temperature.
• Provides information on adsorbed species and on catalyst thermal stability	— Oxidation/Reduction.
• Inert (N_2) or reactive (air/O_2) atmospheres.	— Other Chemical Reaction.
	— Phase Changes.
• Useful for determining catalyst poisoning species.	• Provides information on reaction of adsorbed species or of catalyst.

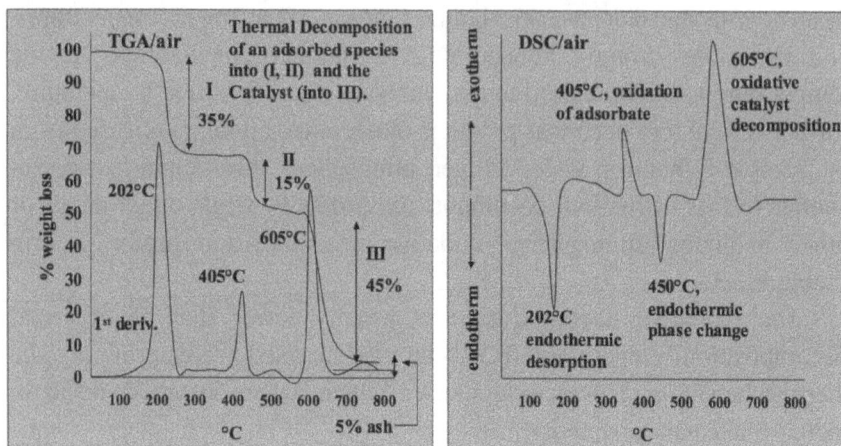

Fig. 7.2.3   TGA/DSC

decomposes at 605°C, makes up 45% of the weight. The 5% of the weight that does not oxidatively decomposed under 800°C is called "ash". The DSC shows that the 202°C desorption processes is endothermic, while the ones at 405°C and at 605°C are exothermic. This suggests that these latter two are oxidative processes, since the experiment was done under air. This could be confirmed if these peaks disappear when the experiment is run under an inert atmosphere.

7.3 Problems

1. a. What spectrocopic method would you use to analyze the surface of an amorphous catalyst?
   b. What method would you use to determine the relative amount of crystalline phases using a polycrystalline sample?
   c. What method would you use to determine the chemical functional groups of adsorbed species on the surface of a catalyst?
   d. What method would you use to obtain a structural image of a catalyst with atomic resolution?
   e. What method would you use to determine bonding distance of nearest-neighbor atoms to a central atom.
   f. What method would you use to obtain an NMR spectrum of adsorbed species on the surface of a solid catalyst?

2. The following graph illustrates how the ratio of Bi:Mo varies as a function of the ratio of Bi:Mo in the bulk for the catalyst system $Bi_2O_3 \cdot nMoO_3$, where n = 1, 2 or 3.

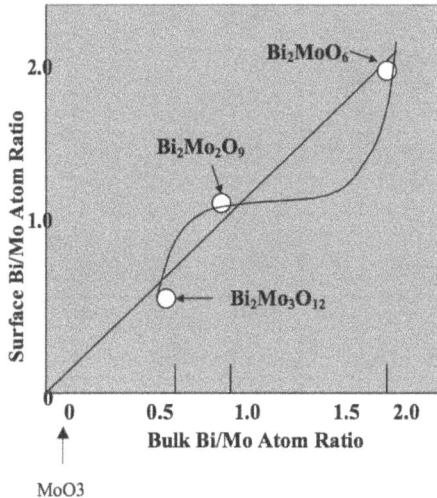

(a) What does this graph say about where Bi goes as the Bi:Mo in the bulk (i.e., the Bi:Mo ratio used to make the catalyst) is varied from 1:1 to 1:1.5?

(b) By what technique could one determine the ratio of Bi:Mo on the surface of these catalysts?

7.4 Answers to Problems

1. (a) XPS or Raman,
   (b) XRD or XRF,
   (c) Raman or FTIR,
   (d) TEM, SEM,
   (e) EXAFS,
   (f) MASNMR.

2. (a) Bi goes into the bulk to maintain the surface at a 1:1 Bi:Mo molar ratio,
   (b) By XPS.

References

1. J.A. Moulijn, P.W.N.M. van Leeuwen and R.A. Santen, eds., *Studies in Surface Science and Catalysis 123*, Catalysis, An Integrated Approach to Homogeneous, Heterogeneous and Industrial Catalysis, Elsevier, New York, (1999).
2. D. Kolb, *J. Chem. Ed.*, **56**, 11, p. 743, (1979).
3. J. Trofast, *Perspect. Catal. [Proc Swed. Symp. Catal.]*, 12th, **46**U, p. 9, (1981).
4. M. Brookhart, M.L.H. Green and G. Parkin, *Proc. of the National Academy of Sciences of the United States of America*, **104**(17), 6908, (2007).
5. B.C. Gates, J.R. Katzer and G.C.A. Schuit, *Chemistry of Catalytic Processs,* McGraw-Hill, New York, 1979.
6. G.W. Parshall, *J. Molecular Catalysis*, **4**, p. 243, (1978); G.W. Parshall, S.D. Ittel, "Homogeneous Catalysis", J, Wiley and Sons, New York, (1992).
7. R.K. Grasselli, J.D. Burrington, D.J. Buttrye, *et al.*, *Topics in Catalysis*, **23**(1–4), pp. 5–22, (2003).
8. M. Misono and T. Okuhara, *Chemtech.*, p. 23, (1993).
9. Shaw and Walker, *J. Am. Chem. Soc.* **80**, p. 5337, (1958).
10. J.W. Ward in *Zeolite Chemistry and Catalysis,* J.A. Rabo, ed., ACS Monograph 171, American Chemical Society, Washington D.C., (1976), pp. 118–284.
11. C.F. Payn, "Solid Acid Technology Impacts to 2000," presented at the Worldwide Solid Acid Process Conference November, pp. 14–16, 1993, Catalyst Consultants Inc, Spring House, PA, (1993).

12. M.L. Poutsma, in *Zeolite Chemistry and Catalysis,* J.A. Rabo, ed., ACS Monograph 171, American Chemical Society, Washington D.C., 1976, pp. 437–528.

13. H.W. Haynes, *Catal. Rev. Sci. Eng.,* **47**, p. 273, (1978).

14. M. Csicsery in *Zeolite Chemistry and Catalysis,* J.A. Rabo, ed., ACS Monograph 171, American Chemical Society, Washington D.C., 1976, pp. 680–713.

15. N. Mizuno and M. Misono, *Chemical Reviews.,* **98**, pp. 199–217, 1998.

16. J.D. Burrington, J.R. Johnson and J.K. Pudelski, *Topics in Catalysis,* **23**(1–4), pp. 175–181, 2003.

17. Y. Izumi, *et al., Bull. Chem. Soc. Jpn.,* **62**, p. 2159, (1989); Y. Izumi, K. Urabe, M. Onaka, "Zeolite, Clay and Heteropolyacid in Organic Reactions," pp. 126–128, VCH, New York, (1992).

18. K. Weissermel and H.-J. Arpe, *Industrial Organic Chemistry,* Second Ed. VCH, (1993), New York.

19. R.K. Grasselli and J.D. Burrington in *Advances in Catalysis,* **30**, p. 133, Academic Press, New York, (1981).

20. R.K. Grasselli, J.F. Brazdil and J.D. Burrington, *Applied Catalysis* **25**, pp. 335–344, (1986).

21. L. Nakka, J.E. Molinari and I.E. Wachs., *J. Am. Chem. Soc.,* **131**(42), pp. 15544–15554, (2009).

22. M.A. Pepera, J.L. Calahan, M.J. Desmond, *et al., J. Am. Chem. Soc.,* **107**, pp. 4883–4892 (1985).

23. M. Abon and Jean-Claude Volta, *Applied Catalysis A: General* **157**, pp. 173–193, (1997).

24. H.A. Wittcoff and B.G. Reuben, *"Industrial Organic Chemicals in Perpsective,"* John Wiley, New York, (1973).

25. J.E. McGrath, ed., *"Anionic Polymerization: Kinetics, Mechanism and Synthesis,"* ACS Symposium Series No. 166, (1981).

26. J.E. McGrath, *"Polymer Synthesis: Fundamentals and Techniques",* May 18–19, (1995).

27. M. Morton, L.J. Fetters, *Rubber Chem. Technol.,* **48**, p. 359, (1975).

28. E.R. Santee, Jr., L.O. Malotky and M. Morton, *Rubber Chem. Technol.,* **46**, p. 1156, (1973).

29. S. Bywater, D.H. Mackerron, D.J. Worsfold *et al., J. Poly. Sci.* Polym Chem Ed., **23**, 1997 (1985).

30. (a) *SRI Competitive Surveillance: The Metallocene Revolution,* 1993; (b) *"Handbook of Transition Metal Polymerization Catalysts,"* R. Hoff and R.T. Mathers, ed., John Wiley and Sons, Hoboken, NJ, 2010; (c) *"Hybrid*

Catalysis in Polyolefin and Copolymers", G. Bazan, W. Brittain, R. Quirck, P.J.T. Tait, K. Wagener, The Catalyst Group, Spring House, PA, (1995).

31. J.C. Floyd, *"Advances in Metallocene Catalysis for Polyolefins,"* Advances in Catalytic Technologies (ACT) Seminar, October pp. 17–19, 1993, Redwood, CA; Hydrocarbon Processing, Int. Ed., **73**(3), pp. 57–60, 62.

32. P. Polanowski, J.K. Jeszka and K. Matyjaszewski, *Polymer,* **55** (10), pp. 2552–2561, (2014); K. Matyjaszewski, *Polymer Preprints (American Chemical Society, Division of Polymer Chemistry*, **38**(2), pp. 383–384, (1997).

33. J. Chiefari, Y.K. (Bill) Chong, F. Ercole, *et al.*, *Macromolecules,* **31**(16), pp. 5559–5562, (1998).

34. G. Morini, E. Albizzati, G. Balbontin, *et al.*, *Macromolecules*, **29**(18), pp. 5770–5776, (1996).

35. W. Keim, *Angew. Chem. Int. Ed.* 12492, (2013).

36. E.F. Lutz, *J. Chem. Ed.,* **63**(30), (1986).

37. L.K. Johnson, C.M. Killian, M. Brookhart, New Pd(II)- and Ni(II)-Based Catalysts for Polymerization of Ethylene and α-Olefins, *J. Am. Chem. Soc.,* **117**, pp. 6414–6415, (1995).

38. D. Ittel, Lynda, K. Johnson and M. Brookhart, *Chemical Reviews*, **100**(4), pp. 1169–1203, (2000).

39. R.H. Grubbs and G.W. Coates, *Acc. Chem. Res.*, **29**(2), pp. 85–93, (1996).

40. J.M. Thomas and W.J. Thomas, *Principles and Practice of Heterogeneous Catalysis*, VCH, New York, (1996).

41. P. Biloen and W.M.H. Sachtler, *Adv. Catal.* **30**, p. 165. (1981), P. Biloen, Recueil, *J. the Royal Netherlands Chemical Society*, **99**(2), (1980). P. Biloen, J.N. Heile and W.M.H. Sachtler, *J. Catalysis* **58**, pp. 95–107, (1979).

42. Retrieved from www.dieselnet.com, 2015 W. Addy Majewski, *DieselNet Technology Guide*, (a) Diesel Emissions, (b) Diesel Oxidation Catalyst, (c) Lean NOx Catalyst, (d) Selective Catalytic Reduction, (e) NOx Absorbers, (f) Diesel Filter Systems, (g) Cellular Monolyth Substrates, (h) Catalyst Fundamentals, (i) Diesel Particulate Filter, Wall Flow Monolyths, (j) Diesel Filter Systems, Catalyzed Diesel Filters.

43. H. Klein, S. Lopp, E. Lox, *et al.*, SAE Paper 1999-01-0109, *Diesel Exhaust After Treatment,* 1999 (SP-1414).

44. Critical Topics in Exhaust Gas Treatment, Research, Research Studies Press, Philadelphia, PA, p. 110, (2000).

45. Fuel Cell Handbook, Fifth Ed., DOE/NEtL-2000/1110. Retrieved from www.netl.doe.gov.

46. T. Ono, N. Ogata, H. Numata, *et al*., *Topics in Catalysis* **15**(2–4), p. 229, (2001).

Index

A

acetic acid, 2, 4, 107, 169, 184,
  187–189
acrolein, 10, 29, 85, 96–97, 99,
  105
acrylonitrile, 4, 35–36, 105, 127,
  133
    acrylonitrile/butadiene/styrene,
      105
    ammoxidation of propylene
      to acrylonitrile, 106
    acrylonitrile-styrene-butadiene,
      76
    propylene to acylonitrile, 34
activation energy, 7, 10, 215
activity, 2, 11–15, 17, 21, 31–32, 45,
  60, 63–64, 68–69, 101, 143, 149,
  150–151, 199, 255
    Raman activity, 246
    Surface Activity, 63
adipic acid, 34, 104, 108
advantages of heterogeneous
  Catalysis, 30–31

Ag, 2, 34–35, 96, 118, 205
    AgO, 36
    Ag = O, 96, 118
    Ag–O, 96
    AgO, 102, 117
    –O–Ag–O, 102
    Cl–Ag–Cl sites, 102
agostic, 24, 141, 161
    β-agostic, 161
Al, 121, 149, 152
    AlCl$_3$, 129, 139, 145, 156,
      162–163
    [AlCl$_3$(OH)$^-$], 145
    [AlCl$_3$(OH)]$^-$, 158
    [AlCl$_3$(OH)]$^-$ H$^+$, 163
    AlEi$_2$Cl, 131
    Al-Et$_2$Cl, 131
    AlEt$_2$Cl, 146, 149, 162–163
    AlEt$_3$, 139–140, 150–151
    AlR$_2$Cl, 147
    AlR$_2$X, 131
    AlR$_3$, 131
    aluminum alkyl, 130–131, 140

[(CH$_3$)C$^+$][AlCl$_3$(OH)$^-$], 145
Et$_2$AlCl, 138
(Me–Al–O–)n, 131
organoaluminum compounds, 121
[R+(AlCl3OH)–], 145
Al, 51–52, 58
  AlCl$_3$, 49, 79, 81–82
  H + [AlCl$_3$OR]$^-$, 82
  AlCl$_3$, 35
  Al$_{3+}$, 51
AlBN, 162
aldol, 187
aldol condensation, 186–187
alkylation, 3
alpha-elimination, 24
alumina, 31–32, 45, 50–51, 63–64,
  68, 81, 178, 200, 208, 220, 246
  Al$_2$O$_3$, 5, 35, 50, 177–178, 209,
    213, 227, 230, 233
  AlO$_3$ alumina units, 51
  AlO$_4$, 58
  aluminum, 249
  BaO/Al$_2$/O$_3$, 205
  dealumination, 249
  high surface area alumina, 210
  γ-Al$_2$O$_3$, 213, 231
  γ-alumina, 205
  platinum-on-alumina, 210
  TiO$_2$ (anatase), 234
  tetrahedral AlO$_4$, 51
ammonia, 3, 19–20, 52, 190,
  201–202, 208, 222–225
  ammonia reduction of NO,
    212
  ammonia synthesis, 169, 184,
    190–193, 196–197
  ammonia synthesis
    (Haber– Bosch), 3, 34

ammonia slip, 225
2NH$_3$, 7
NH$_3$, 18, 35–36, 105, 117, 163,
  196, 203, 208, 223–224,
  226, 231, 233–234
  reduction by ammonia, 209
  reduction of NO with ammonia,
    203
  oxidation in the presence of
    ammonia, 105
  thermodynamics
    (NH$_3$-decomposition), 3
ammoxidation, 105
  ammoxidation of propylene,
    105
amorphous, 52, 150–152, 213, 239,
  241–242, 246, 256
anionic polymerization, 127–128,
  145, 156
aromatic, 58, 72, 82, 176, 189
  alkyl aromatic, 55, 59, 81–82
  aromatic alkylation, 67, 76
  aromatic Reactions, 55
  aromatic oxidation, 84, 104,
    110–111
  polynuclear aromatics, 221
Auger, 238

B

(Ar$_4$B)$^-$, 159
BF$_3$, 129, 144, 157, 162
[BF$_3$(OH)]$^-$, 144–145, 158
BF$_3$(OR)$^-$, 144
BF$_3$/ROH, 145
B, Al, 120
BF$_3$, 35, 76–77, 79
(C$_6$F$_5$)$_4$B$^-$, 144
(CH$_3$)$_3$C$^+$, 144
[(CH$_3$)C$^+$][BF$_3$(OH)$^-$], 144

[H$^+$][BF$_3$(OH)$^-$], 144
Ph$_3$C$^+$BF$_4^-$, 129
[–Si–O–BF$_3$]$^-$ H$^+$, 78
[–Si–O–PO$_3$–BF$_3$]$^-$ H$^+$, 78
Ba, 35, 200, 205, 226, 232
Ba(NO$_3$)$_2$, 209, 212, 226, 230, 233
BaO, 209–210, 213, 226–227,
    230–231, 233–234
barium oxide, 209
beckman isomerization, 78
β-elimination, 24, 94, 161, 165, 186
benzene, 34, 71, 76–77, 87–88,
    189–190, 192, 196
benzoic acid, 94
Berzelius, 6
BET, 253
Bi, 34, 96–97, 257
    Bi (2+), 96
    Bi (3+), 96
    Bi$_{3+}$ or 5+, 97
Bi:Mo, 29–30, 102, 257
bismuth molybdate, 29, 96
    Bi$_2$Mo$_3$O$_{12}$, 29–30, 96, 105,
        110, 115, 117–118
    Bi$_2$MonO$_{3+}$3n, 97
    Bi$_2$MoO$_6$, 117
    Bi$_2$O$_3$•nMoO$_3$, 101, 257
    Bi$_2$Mo$_2$O$_9$, 101–102
    Bi$_2$Mo$_3$O$_{12}$, 101
    Bi$_2$MoO$_6$, 101–102
    α–bismuth molybdate, 29, 105
Bragg equation, 242
branched Polyethylene, 159
bronsted acid, 44–45, 48–50, 53,
    76, 78, 82, 129
    bronsted acid-catalyzed, 67
    bronsted acid site, 55, 71, 76–77
    bronsted sites, 53, 57–58,
        66–67

Brookhart, 5
Brookhart catalyst, 158
β-scission, 69–70
butadiene, 102, 105, 110, 127, 129,
    133, 145–146, 154–155, 247
butadiene and styrene, 154
butene, 102, 152, 247
butyraldehyde, 26, 186–187

C

(CH$_3$AlO)n, 163
calcination, 53–54, 82, 178
carbenium, 71
carbenium ion, 55, 79, 126, 190
    1° carbenium ion, 70
    2° carbenium ion, 69–71,
        190
    2-hexyl carbenium ion, 70
    alkyl carbenium ion, 55
    cyclohexadienyl carbenium ion,
        55
    tert-butyl cation, 72–74
carbon, 32
catalyst life, 21, 101
catalysts (Pt/Rh/La), 5
catalytic mechanism, 21, 25
    catalytic mechanism example,
        25
catalytic reforming, 184, 189–190,
    192–193, 196–197
Ce, 97, 200, 210
    Ce$_{3+}$/Ce$_{4+}$, 97
    CeO, 35, 205, 212–213, 227,
        230–231, 233–234
    CeO$_2$, 227
    CeOx, 222, 227
    Ceria, 205
    CH$_3$(C = O)RhI$_3$(CO)$_2$,
        196

chain termination, 75, 133, 135, 140, 148

chain transfer, 70–75, 122, 127–129, 132–137, 147–148, 154–156, 165, 176, 179

chemisorption, 3, 8, 23–27, 30, 185–186, 237

Cl, 102–103
    Cl$^-$, 107, 132
    Cl$_2$, 102
    Cl–Ag–Cl sites, 102
    Cl promoter, 102

Clay, 31–32, 45, 50, 68
    acidic clay, 79

Co, 23, 31, 34, 107, 170, 178–181, 194, 200
    Co catalyst, 177, 179–180, 186–188
    CoI$_2$, 187–188
    Co$_2$(CO)$_8$, 25
    Co/PPh$_3$, 194
    Co/SiO$_2$, 174–175, 191–192, 196
    Co (2+), 94, 108, 117
    Co (3+), 94–95, 108, 117, 110
    Co–naphthenate, 117
    Co–naphthenate catalyst, 115, 117
    HCo(CO)$_4$, 25, 38, 41, 187–188
    HCo(CO)$_3$, 25–26

Co/Mo/S/Al$_2$O$_3$, 177–178

coke, 10, 56, 59–61, 71, 82, 122, 171, 184

combustion, 10, 33, 56, 83, 202, 204, 216, 229
    catalytic combustion of hydrogen, 2
    combustion catalyst, 218

internal combustion engine, 228

internal combustion, 177

contact time, 36–37, 40

coordinately saturated, 24

coordinately unsaturated, 23, 26–29, 38

Cope elimination, 23

Cr, 121
    CrO$_3$, 131

cracking, 3–4, 55, 81–82
    catalytic cracking, 3, 69–70, 73, 76
    cracking catalyst, 60
    fluid catalytic cracking, 34
    hydrocracking, 169
    hydrocarbon cracking, 69
    isobutane cracking, 57

crystalline, 4, 150–152, 241–242, 246, 256
    polycrystalline, 239–240, 256

crystalline solids, 27

Cu, 35, 105, 107, 170, 200, 205, 208, 213
    Cu$^+$, 94
    Cu(1+), 107
    Cu$_{2+}$, 94, 107
    Cu$_{2+}$/Cu$^+$ couple, 94
    Cu$_2$O$_3$, 99
    CuCl$_2$, 117
    CuO$_x$ catalysts, 99
    Cu for NO oxidation, 212
    Cu/ZSM5, 35, 205

cumene, 34, 76, 78, 104, 111–112

cyclohexane, 34, 108, 189–190, 192, 196
    cyclohexane 1,2-diol, 88
    cyclohexane oxidation, 107, 110

cyclohexanol, 104, 107–108
    cyclohexanone oxime, 78, 107
cyclooctadiene, 159

D

4,4-dimethyl 1-hexene, 81–82
4,4-dimethylhexene, 60
DCS, 255
dealkylation, 55
decarboxylation of malonic acids, 23
decomposition of H_2O_2, 2
dehydration, 33, 57, 74–76, 98
dehydration of alcohol, 2
dehydrogenation, 34, 176, 184–186, 189, 192, 196–197, 221
Diels–Alder reaction, 23
diesel oxidation catalyst (DOC), 35, 204–205, 210, 219–222, 225–226, 230, 233
diesel particulate filter (DPF), 35, 204–205, 210, 216, 219–221, 230, 233
differential scanning calorimetry (DSC), 237, 252
di-isobutyl phthalate, 150
disproportionation of aromatics, 71, 72
donor molecules, 150

E

2-ethylhexanol, 187
β-elimination, 153, 157
EDAX, 251
electron energy loss, 251
electron microscopy (EM), 250
electron Pushing, 22
ene reaction, 23

engines, 177
enzymes, 2, 5
epoxidation, 28, 84, 92–93, 95, 102, 104
    ethylene epoxidation, 34
epoxide, 29
EPS, 243
ethylbenzene, 34, 76, 105, 184–186
ethylene, 4, 13, 15–16, 34, 37, 39–40, 76, 94–96, 102, 107, 150–154, 158–161
    ethylene oligomerization, 35
        heterogeneous oxidation of ethylene, 36
ethylene oxide, 36–37, 39, 95–96, 102
EXAFS, 244, 257
extended X-ray adsorption fine structure, 238

F

Fe, 34, 121, 170, 178–179, 191, 197, 200
    Fe_2O_3, 191, 195
    Fe_3O_4, 196
    Fe catalyst, 179
    Fe/SiO_2, 174–175, 191, 196
    Fe_{2+}/Fe_{3+}, 97
    $Fe_aSe_bTecO_x$, 97
    $Fe_XSb_YO_Z$, 97
fermentation, 2
Fischer–Tropsch, 3, 75, 175–178, 191–192, 196
formed, 26
Fourier transform, 245
Fourier transform infrared, 238
Fourier transform infrared (FTIR) spectroscopy, 245, 257

Fourier transform NMR (FTNMR), 248

Free radical, 22, 90, 92, 104, 110–111, 115, 121, 125–126, 133–134, 155, 163–164
    benzylic peroxide radical, 105
    benzyl peroxide radical, 94
    benzyl radical, 94–95, 104
    cyclohexyloxy radical, 108
    cyclohexyl radicals, 108
    organic radical, 90
    stable free radical, 136

H

1-hexene, 60, 81–82, 152
Haber–Bosch, 169, 184, 190–191
Hammett acidity scale, 46
heterogeneous catalysis, 1, 26–27, 29–30, 32, 253
    advantages of homogeneous catalysis, 30
    advantages of heterogeneous catalysis, 31
    heterogeneous catalysis example, 29
    homogeneous Catalysis example, 28
heterogeneous oxidation, 90–92, 95, 101, 110
    H-abstraction, 91, 96–97, 105
    homogeneous free radical, 90, 94
    free radical homogeneous, 107
heteropolyacid, 47, 64, 98, 157
    $Cs_{2.5}H0$, 65
    $CsxH_3–xPW_{12}O_{40}$, 64
    $H_{0.5}(NH4)_{2.5}PW_{12}O_{40}$, 145
    $H_3+xPW_{12}-xVxO_{40}$, 98

$[(NH_4)_{2.5}PW_{12}O_{40}]^{-\frac{1}{2}}$, 158
hexamethylene diamine, 109
high surface area, 213
Ho Acidity Scale, 46
hydride abstraction, 57, 70–71, 73
hydrocarbon oxidation, 211
hydrocracking, 184–185, 192, 196–197
hydrodesulfurization, 4, 177
hydrodesulfurization (Co/Mo/S), 3
hydroformylation, 4, 23–26, 34, 172–173, 180–182, 184, 186–187, 192, 194, 196
hydrogenation, 10, 33, 60, 82, 169, 170, 172, 178, 180, 183–184, 187, 189–191, 193
hydrogenolysis, 56, 122, 169, 171, 184–185
hydrogenolysis/Methane, 10
hydrolysis, 33, 226
    hydrolysis of urea, 7
    hydrolysis of esters, 9
    urea hydrolysis, 208
hydroperoxide, 29, 105
    benzylic hydroperoxide, 105
hydrosulfurization, 169
hydrotreating, 169, 184–185, 192, 196–197
hydrotreating/hydrocracking, 34

I

initiation, 125, 127, 129, 133–136, 165
    initiator–polymer complex, 135
    SDS, 154–155
initiator, 71–73
    TEMPO Initiator, 136

ion exchange resins, 50
    Amberlyst, 4
    polystyrene Amberlyst, 50
isobutane, 72–74, 80
isobutylene, 74, 129, 144–145
isomerization, 33, 55, 57, 61, 70–71,
    81–82, 157, 160, 189, 195
    Beckman isomerization, 77,
        78, 80
    hydroisomerization, 4
    paraffin isomerization, 34
isooctane, 72–73
isotactic, 143

K

keto–enol tautomerism, 23
kinetics, 2, 7
KNH_2, 163
KOH, 162

L

La, 200
La_2O_3, 213
lanthanum oxide, 205
LaO, 35, 205, 213, 227, 230–231,
    233–234
late transition metal, 120, 149,
    158–159
Lewis acid, 32, 44, 49, 53, 57, 76, 79,
    81–82, 129–130
    Lewis acid catalysts, 49
    Lewis acidity measurements
        in solids, 53
    Lewis acid site, 53, 67, 69
    Lewis sites, 57, 70
Lewis acidity, 48, 54
Lewis base, 22, 149–150
life, 2, 11, 71, 199, 212

ligand dissociation, 23, 27, 30, 38,
    41, 105

M

2-methylpentane, 189–190
magic angle spinning nuclear
    magnetic resonance (MASNMR),
    248–249, 257
magnesia, 68, 211
    platinum-and-rhodium-on-
        magnesia catalyst, 210
maleic anhydride, 104, 109, 111,
    133
MAO, 133, 144, 163
Mars–van Krevelen, 90–91, 95–97,
    105, 110, 115, 117
mercaptan, 155
metallocene, 121, 130, 132–133,
    141–144, 147–148, 159, 163–164,
    172
metathesis, 3
methanation, 173–174, 191, 196
methane, 11, 56–57, 122, 171, 173
methanol, 4, 59, 74, 81, 173, 175,
    187–189
    CO to methanol, 89
    carbonylation of methanol, 169,
        184, 187–188
    methanol carbonylation, 192,
        196
    methanol decomposition, 174,
        191, 196
    methanol synthesis, 174,
        191–192, 195–196
    Methanol-to-Gasoline, 34, 74
Mg, 35, 200
    Me_2Mg, 162
    $MgCl_2$, 34, 131, 139, 146–147,
        149–151, 162–163

MgO, 212–213, 216, 231, 233–234
migratory insertion, 24, 26, 38,
    139–140, 142
Mo, 29, 34, 47, 92, 96, 104, 115, 121,
    177, 245–246
    Bi_{2+}–OH, 118
    Bi_{3+}–O, 118
    MoS_2, 245
    Mo6+=O, 30
    MoL_2O_2 (OH)(OR), 28
    Mo-oxo complexes, 28
    MoO_3, 49
    Mo(4+), 96
    Mo(5+), 96
    Mo(6+), 96
    Mo(acac)2, 117
    Mo = NH species, 105
    Mo = O, 96, 105
    Mo–O–OR, 117
    Mo oxo species, 105
    Mo–peroxide complex, 105
mordenite, 61
MTBE, 4

N

n[$CH_3Al(Cl)$–O–Al(Cl)–O]$^-$, 163
Nafion, 50
Ni, 34, 149, 153, 159–161, 170, 200,
    246
    $Ni(COD)_2$, 162
    $NiL(CO)_3$, 181
    NiO/SiO_2, 174
    NiOx, 196
    Ni/SiO_2, 191, 195
    organo-Ni catalyst, 153
Ni/Mo, 35
NMR, 247–249, 256
NO oxidation, 211

NO, 200–201, 203, 205–210,
    212, 218, 221–226, 228,
    230–235
    NO_2, 201, 203, 205, 208–210,
        212, 218, 221, 226, 228,
        230, 233, 235
NO_x, 33, 35, 199, 201–205,
    207, 209–210, 212–213,
    218, 222, 224, 226–228,
    230–231, 234
    3-way catalyst, 205
    Lean NOx, 35, 230
    Lean NOx Catalyst, 205, 207
    N_2O, 208, 222–223
    NH_3 reduction of NO, 213
    NO oxidation, 213
    NO_x absorber, 205, 209–210,
        232, 230, 233
    NO_x decomposition, 205,
        212–213
    NO_x Trap, 226–227
    NO, 246
    NO_x Absorber, 35
nuclear magnetic resonance, 238
nylon 6,6, 108–109

O

1-octene, 152
octahedra, 47
O-donor, 150
olefin copolymer
    OCP, 153, 162
olefin copolymers, 120, 149, 152
oxidation, 104
    CO oxidation, 211
oxidation of propylene, 29
oxidation of propylene to acrolein,
    29

oxidation state, 27, 33, 44, 48, 83–85, 87–89, 101, 120–121, 170, 243, 244, 251
    organic oxidation states, 85, 87–89
    relative oxidation states, 86
oxidative addition, 24, 26
oxo alcohols, 4, 26, 182, 186–187
oxydehydrogenation, 247

P

$5PW_{12}O_{40}$, 65
Pd, 4, 35, 105, 107, 159–160, 170, 184, 200, 205, 211, 219, 222
    Pd0, 94
    Pd^{2+}, 94
    Pd (2+) di-imine, 159
    Pd(COD), 162
    Pd/Cu, 34, 94
PEM, 228, 232
peroxide, 90, 108
    alkyl peroxide, 92
    benzyl peroxide, 95
    di-cyclohexyl peroxide, 108
    hydrogen peroxide, 92
    peroxide initiator, 104
    peroxides decomposition, 104
Ph_3C^+, 144
phenol, 76, 111–112
phenyl triethoxysilane, 150
Phillips catalyst, 130
phosphine, 31, 182
    phosphine ligand, 180–182
    triphenyl phosphine, 182
    tris(2-methylphenyl) phosphine, 182
phosphotungstic acid, 46–48, 66

Phthalic anhydride, 104, 111–112
pKa Acidity Scale, 45–46
PM, 202–203, 210
    particulate material, 210
poison, 3, 10, 56, 169, 177–178, 184, 199, 212–213, 219
poisoning, 255
polybutadiene, 127–129
    Krayton® rubber, 127, 162
polybutadienes, 145
polydispersity, 123–125, 161
polyethylene, 120, 130–132, 149–153, 161
    comonomers for polyethylene, 153
    high density polyethylene (HDPE), 152
    linear low density polyethylene, 152
    low density polyethylene (LDPE), 152
    metallocene polyethylene mechanism, 153
    properties of polyethylene, 152
polyethylene, 3, 16, 34, 150
    branched polyethylene, 158, 162
    polyethylene branching mechanism, 161
    perfluorinated sulfonated polyethylene resin, 50
polyethylene catalysts, 159
polyethylene, polypropylene, 119
polyisobutylene, 120, 149, 159
    β-isomer, 156–157
    high vinylidene PIB, 157
    tri-and tetra-substituted, 156

two types of products:
  conventional PIB, made
  from AlCl3 catalysts, which
  is comprised of mainly
  tri- and tetra-substituted, 156
  vinylidene, 158
  vinylidene isomer, 145
polyisobutylene, 73, 129–130, 156
  conventional (Trisubstituted)
    polyisobutylene, 162
  high vinylidene
    polyisobutylene, 162
polymerization
  atom transfer radical
    polymerization, 137
  chain-growth polymerization,
    135
  free radical emulsion processes,
    149
  free radical polymerization,
    133–134, 136
  living free radical
    polymerization, 135
  living polymerization,
    135–137
  polymerization catalyst,
    119–122, 129, 137, 143, 146
  reversible addition-
    fragmentation chain
    transfer, 137
polymerization catalysts, 31, 33
polynuclear aromatic compounds,
  82
polypropylene, 34, 141–142, 149–150
  amorphous polypropylene, 150
  isotactic, 4, 139–142, 146,
    149–150
  isotactic polypropylene, 3, 162
  syndiotactic, 139, 145

syndiotactic polypropylene, 162
V-Based PP catalysts:
  syndiotactic, 140
polypropylene oxide, 162
polysobutlyene
  vinylidene, 156
polystyrene, 34–35, 50, 76, 119, 155
pore size, 27, 63–64, 237, 252, 254
pore volume, 63–64, 237, 252, 254
potassium persulfate initiator, 154
promoter, 3, 83, 96–97, 102, 169,
  178, 184, 199, 211–213, 231, 234
  Oxidation catalyst promoters, 103
propagation, 126–129, 134, 136,
  147–148, 154, 165, 176, 179,
  194
propylene, 3, 10, 28–30, 34, 69–71,
  76, 96, 99, 105, 140, 142–143,
  150–152, 187, 194
  polymerization of propylene,
    138
  propylene ammoxidation, 4
  propylene oxidation, 4
  propylene to propylene oxide,
    105
  propylene Copolymers,
    150–151
  propylene Polymerization, 151
  selective oxidation of
    propylene, 96
  syndiotactic, 138
  total oxidation of propylene,
    85
propylene oxide, 4, 104–105
Pt, 2, 81–82, 170, 184–185, 189, 198,
  200, 205, 209–211, 213, 216,
  219–222, 225–230, 232–234
  Pt/Al$_2$O$_3$, 35, 81, 230, 233
  Pt–C, 228

Pt–H, 229
Pt–H species, 228
Pt metal catalyst, 193
Pt metal, 208
Pt-NO$_2$ species, 208
Pt-O, 228, 229
Pt on Al$_2$O$_3$, 225
Pt on C, 228
Pd (0), 107
Pd(2+), 107
PdCl$_2$, 117
[PdCl$_2$]$_2^-$, 107
[PdCl$_4$]$_2^-$, 107
Pt-based exhaust gas catalyst, 246
Pt/zeolite, 196

R

Raman, 238, 257
reactions, 67
reduction, 5, 24, 56, 85–86, 89, 91, 96, 99, 107, 169–170, 178, 184, 199, 201, 203, 205, 207, 228, 255
    NO$_x$ reduction catalysts, 177
    NO reduction, 207, 209, 211, 234
    NO$_x$ reduction, 203, 209, 222, 225–226
    reduction of NO, 210, 227–228, 223, 246
reductive elimination, 24, 26, 172, 179, 183, 185
reforming, 3, 195
relative oxidation states, 87
residence time, 12–13, 16, 37, 40
Rh, 4, 23, 31, 34–35, 170, 172, 180, 182–183, 186, 188–200, 205, 210–213, 216, 226–228, 230–234

CH$_3$(C = O)RhI$_3$(CO)$_2$, 192
CH$_3$RhI$_3$(CO)$_2$, 192, 196
[C(= O)CH$_2$CH$_2$R], 196
[C(= O)CH$_2$CH$_2$R], 192
HRh(CO)$_3$I$_2$, 188
HRh(CO)L$_2$, 196
HRh(CO)L$_2$(CH$_2$ = CHR), 192
HRhI$_3$(CO)$_2$, 192, 196
Rh(CO)$_2$L$_2$, 192, 196
Rh(CO)$_2$L$_2$(CH$_2$CH$_2$R), 192, 196
Rh–H, 172, 183
RhH(CO)$_3$I$_2$, 192, 196
RhH(CO)[PPh$_3$]$_2$, 192, 196
RhI$_2$, 188
RhI$_3$(CO)$_2$, 196
Rh/PPh$_3$, 194
Rh(PR$_3$)$_3$, 195
Rh/SiO$_2$, 174
Rh–C, 227
Rh–N, 227–228
Rh–O, 227–228
rhodium, 210
rhodium nitride, 212
Rh on alumina, 243
Ru, 35, 205

S

Sb, 100, 101
Fe$_X$Sb$_Y$O$_Z$, 97
Sb$_{3+}$, 97
Sb (5+), 100
Sb$_{5+}$, 97
SbF$_5$, 46
scanning electron microscopy, 238
scanning TEM, 238
scanning tunneling microscopy, 238
Schultz–Flory, 169, 176, 193

SDS, 163
Se
    $Fe_aSe_bTecO_x$, 97
    Se_{4+}, 97
secondary ion mass spectrometry
    (SIMS), 239, 251–252
selective catalytic reduction (SCR),
    35, 202–203, 205, 208, 212, 222,
    224–226, 230, 233
        SCR catalyst, 226
        SCR components, 225
        SCR mechanism, 223
        silica, 200
        SiO_2, 213
selective oxidation, 4
selectivity, 2, 11, 17–18, 20–21, 32,
    36–37, 39, 68, 80, 96, 99, 102,
    120, 122, 138, 141, 143, 149–151,
    178, 181, 194
        stereoselectivity, 139
        selectivity example, 18–19
        product shape selectivity, 61
        reactant selectivity, 60
        restricted transition state
            selectivity, 61
        Shape Selectivity, 60
SEM, 250, 257
shape selectivity, 62, 75, 81–82
shell higher olefin process, 4, 153
SHOP catalyst, 159
Si, 51
silica, 31, 32, 45, 50–51, 63–64,
    68–69, 76, 131, 246, 249, 252
    SiO_2, 50, 104
silica–alumina, 4, 34, 45, 50–52, 55,
    68, 74, 79, 81–82, 189, 249
        crystalline silica–alumina, 50,
            52, 58
        amorphous silica–aluminas, 32

        crystalline silica–aluminas, 32
        SiO_2/Al_2O_3, 51
        crystalline silica–aluminas, 50
silver catalyst, 95, 115, 118
site density, 15–16
SMPO process, 104
SN_2 substitution, 23
sodium dodecylbenzene sulfonate,
    154
soot, 33, 35, 199, 202–203, 210–212,
    216–221, 226, 228
soot balance point, 217
soot light off temperature, 217
space velocity, 12–16
        gaseous hourly space velocity,
            12–14
        liquid hourly space velocity,
            12–13
        space velocity examples, 13
        volumetric space velocity, 12
        Weight Space Velocity, 12
steam reforming, 171, 173–174, 191,
    195, 228
STEM, 251
stereoselectivity, 141, 183
stress cracking, 152
styrene, 34, 76, 104, 129, 134, 136,
    138, 154–155, 184–186
        copolymerization of styrene,
            137
        methacrylate/styrene block
            copolymer, 137
        polymerization of styrene, 125
        SBR, 120, 155
        styrene/butadiene emulsion
            polymerization, 154–155
        styrene butadiene rubber, 76,
            149, 162
styrene/butadiene, 156

succinamide, 18
succinic anhydride, 18, 20
surface activity, 64–65
surface area, 63–64, 68–69, 213–214,
  237, 252
    high surface area support, 209,
      231, 234
synthesis gas, 169, 171, 173–174,
  175–176, 180

T

Te, 121
    $Fe_aSe_bTecO_x$, 97
    Te_2MoO_7, 97
    Te_{4+}, 97
    Te_{6+}, 97
telechelic polymers, 135
TEM, 250, 257
temperature, 98
TEMPO, 136, 137
terephthalic acid, 4, 34, 104,
  110–111
termination, 131, 165
tert-butyl cation, 144
TGA, 252, 255
thermodynamics, 2, 3, 7–8, 10, 56,
  73, 83, 120, 122, 171
    exhaust thermodynamics, 202
    Oxidation thermodynamics,
      85–86
    Oxidation Thermodynamics, 85
thermogravimetric analysis, 237
three-way catalyst
    Three-way, 5
    three-way (NOx-, hydrocarbon-
      and CO-reducing) catalysts,
      5
Ti, 121, 139, 149, 152
    M-Ti, 131

Ti alkyl, 146–147
Ti-based PP catalysts, 140–141
$Ti-CH_2CH(CH_3)-R$, 140
$TiCl_3$, 15–16, 34, 131, 138,
  146–147, 163
$TiCl_4$, 131, 139–140, 149, 151,
  162
Ti–R, 140
$(TiCl_4)$, 3
TiO_2, 104
titania, 31–32, 68, 209, 246
    anatase, 209
    TiO_2, 105, 224–225
    TiO_2 (anatase), 213, 231
    vanadia–tungsten oxide on
      titania, 208, 223
    $\gamma-Al_2O_3$, 234
Tolman formalism, 25
Tolman steric parameter, 181
toluene, 71, 94
    2-ethyl toluene, 61
    acylation of toluene, 67
    ethyl toluene isomers, 63
    oxidation of toluene, 94
    toluene alkylation, 61
    K/toluene, 162
transalkylation, 55, 61
transmission electron microscopy,
  238
trimethylbenzene, 64
turnover, 14–17, 21
    turnover example, 15

U

U, 99–100
    U (5+), 99
    U_{5+}, 97
    U_{6+}, 97
U/Sb mixed metal, 100

U/Sb mixed metal oxide, 100
urease enzyme, 7
 USb_3O_{10}, 97, 100–101
 $USbO_5$, 100–101
 USbOx, 99–100

V

VIN, 35, 98, 121, 131, 139–140, 152, 200, 205, 212–213, 246
 $V(acac)_2$, 139
 V alkyl complex, 140
 V-based PP catalysts:
  syndiotactic, 140
 V-based Z–N catalysts, 152
 $V–CH(CH_3)–CH2–R$, 139
 VCl_3, 131, 140
 VCl_4, 139
 $VCl_4/AlEt_3$, 162
 $VO(OR)_3$, 162
 V_2O_5, 205, 212, 216, 223–225, 231, 234
 V (4+), 224
 V(5+), 223
 V–NH–N=O, 223
 V=O, 223
 V–Ob, 247
 $VO (PO_4)$, 247
 V(4+), 110
 V(5+), 108, 110
 $(VO)_2P_2O_7$, 110, 117
V (4+), 223
V(5+)=O, 223
V/P, 34
V/Ti, 34
V_2
 V_2O_5, 213
vacancies, 23, 27, 29, 91, 96
V-based PP catalysts, 139

vinyl acetate, 4, 34, 105–107
vinyl acetate/Wacker oxidation, 104

W

3-way catalyst, 203–204, 206, 210, 215, 227, 230–231, 233
 3-Way Gasoline Catalyst, 35
W, 47–48, 98, 121
W_3O1_3, 47
Wacker oxidation, 84, 92–93, 105–106
Washburn equation, 254
water gas shift, 171, 173, 191, 195, 204
WO_3, 213, 231, 234
WO_6, 47
WWH, 12–14, 16, 21, 37, 40

X

XPS, 257
X-ray diffraction, 238
X-ray fluorescence, 238
XRD, 240, 250, 257
XRF, 240, 257

Z

zeolite, 4, 31–32, 45, 50, 52–53, 58, 60, 62, 68, 71, 78, 189, 198, 208–209, 225, 231, 234, 251
 ammonium zeolites, 249
 acid sites in Zeolites, 54
 faujacite, 58–59, 67
 mordenite, 50, 59–60, 67, 71, 81
 Pt/zeolite, 191
 Pt/ZSM-5, 60, 82
 Si/Al frameworks, 59–60

sodalite, 58–59
ZSM-5, 4, 59–60, 67, 71, 74, 81–82, 251
zeolite A, 59
Zeolite Acid Sites, 52
zeolite anion, 72
zeolite "cage", 58
zeolite cavity, 61, 82
zeolite frameworks, 59
zeolite structure, 71
zeolite superstructure, 58
Zeolite synthesis, 52
Zeolite cracking catalyis, 4
Zeolite acid sites, 53
Ziegler–Natta, 3, 15, 121, 130, 132, 139, 141, 146–147, 149, 152, 159, 163–164, 172
Ziegler–Natta catalylst, 150
Zn, 170
    Cu/Zn/SiO$_2$, 175
    Zn/Co/SiO$_2$, 174

Zn/CuOx/SiOx$_2$, 195
Zn/Cu/SiO$_2$, 192
Zr, 121, 130, 133, 142, 144, 148, 252
    C–H–Zr bond, 142
    [Cp$_2$Zr(CH$_2$CH$_2$CH$_3$)]$^+$, 164
    Cp$_2$Zr(CH$_3$)$^+$, 133
    Cp$_2$Zr(CH$_3$)$_2$, 141
    [Cp$_2$ZrCH$_3$(CH$_2$ = CH$_2$)]$^+$, 164
    Cp$_2$ZrCl$_2$, 144, 163
    n[Cp$_2$ZrCH$_3$], 163
    nCp$_2$ZrCl$_2$, 163
    rac-Cp*2ZrCl$_2$, 141
    Zr-based metallocene, 141
    ZrCl$_2$, 132
    Zr$^+$ complex, 132
    ZrCp$_2$Cl$_2$, 131
    Zr-metallocene, 154
    Zr+ metallocene, 144
    Zr/SiO$_2$, 252
    Cp$_2$ZrCl$_2$, 34
ZrO$_2$, 213, 252